GUIDEBOOK TO PECTEN SHELLS

RECENT PECTINIDAE AND PROPEAMUSSIIDAE OF THE WORLD

GUIDEBOOK TO PECTEN SHELLS

RECENT PECTINIDAE AND PROPEAMUSSIIDAE OF THE WORLD

A. ROMBOUTS †

Photographs by P.L. van Pel

Edited and Revised by

Henry E. Coomans*
Henk H. Dijkstra*
Robert G. Moolenbeek*
Peter L. van Pel*
* Zoological Museum, University of Amsterdam

ROBERT HALE • LONDON

Frontispiece
Top left: *Chlamys ecoliniana*
Top right: *Semipallium dianae*
Bottom left: *Gloripallium pallium*
Bottom right: *Gloripallium speciosum*

A CHP Production

Published in Great Britain by
Robert Hale Limited
Clerkenwell House
Clerkenwell Green
London EC1R 0HT

ISBN 0 7090 4831 9

Printed in Hong Kong by Colorcraft Limited

CONTENTS

Ton Rombouts, favourite shelling area

FOREWORD

Antonius Rombouts, "Ton" to his friends, was born in Stompwijk, the Netherlands, on 23 September 1917. He received his education in Holland. In 1943 he married Anna E. Bras, they had five children.

Rombouts was consultant for an engineering firm, which company sent him in 1949 to South Africa, where he became managing director. The company transferred him to Brisbane, Australia, in 1961. Later he started his own business for "Scale model making" and "Artistic Consultants and Services". Ton Rombouts passed away on 19 June 1985 at Brisbane, at the age of 67.

In his late teens Rombouts started collecting shells in Holland. When in South Africa he joined the Conchological Society of Southern Africa (October 1953), and he was appointed Honorary Curator of Molluscs at the Transvaal Museum in Pretoria. Whilst living in Australia Ton became a member of the Malacological Society of Australia (Brisbane Branch) in April 1963. He served as Librarian, Secretary and Treasurer at various stages. During 1977-1978 he gave a series of lectures. In his capacity as Member of the Executive Committee he was one of the chief organisers of the Convention of the Malacological Society of Australia held in Hamilton, May 1985.

He travelled extensively overseas, and purchased and exchanged shells all over the world. He owned a collection of worldwide molluscs, his main interest were the Pectinidae. In 1958 he exchanged South African molluscs with the Zoological Museum in Amsterdam, and more recently he sent Australian Conidae to the Museum. In 1981 he became a member of the Netherlands Malacological Society, in its journal "Correspondentieblad" (no.200, 1981, p.1153) he wrote an article on Melo georginae from Moreton Bay. He also published in various Australian journals.

Fore several years Ton Rombouts worked on a book about Pectens. He did not live to see it published as he passed away suddenly, just before the manuscript was about ready for publication. Because of his long connection with the Institute of Taxonomic Zoology (Zoological Museum) of the University of Amsterdam, its Department of Malacology offered to edit and revise the manuscript. The taxonomy was checked by Henk H. Dijkstra, and the final draft typed by Ton's friend Peter L. van Pel, both Honorary Associates of ZMA. Dijkstra also prepared the Appendix, and Collection-manager Rob G. Moolenbeek checked the References. I am grateful to these colleagues for their efforts.

As a token of appreciation Mrs. Rombouts donated the Pecten collection of her late husband to our Museum. She and Mrs. Lil Ratcliffe also supplied the biographical information for this foreword.

Finally Peter van Pel made the photographs, for which he used specimens from several collections. The extremely rare, and the very small scallop species are not illustrated. From the deepwater Propeamussiidae, which are difficult to obtain by collectors, only a selection was photographed.

We are proud and thankful that Ton Rombouts' work is being published posthumously in his native country, one hundred years after Küster & Kobelt finished their monograph on the Pectinidae in 1888.

Dr. Henry E. Coomans
Chairman Dept. Malacology
Zoological Museum Amsterdam

Popular collecting region, Cannonvale Beach, northern Queensland

INTRODUCTION

This study aims to be a guide to the living species of the Pectinidae and Propeamussiidae. The name Pecten is only applied loosely, since shells of this type and general shape are commonly referred to as such, notwithstanding that scientifically they may bear different names. The reason for writing this study is that there as yet is no other publication available as a reference for collectors and others interested in this group of marine shells.

As a collector I experienced for many years the lack of a compact and comprehensive source of information on the subject. Correlated knowledge on Pectens could be obtained, (and then only partly), by consulting numerous general and regional shell books or articles published in scientific journals. This is in contrast to many other molluscan families which have been specifically dealt with in monographs in the past and in popular books in recent years, e.g. Conidae, Cypraeidae, Muricidae, Mitridae, Olividae and Volutidae.

A popular shell like the scallop should have received more attention. I cannot think of any other shell more widely applied for artistic or architectural purposes through the centuries, thereby reminding us of its grace and beauty. Not that this group of molluscs has been scientifically neglected. In fact it was by study of the taxonomy and nomenclature that I realised that, no doubt with the best of intentions of the respective workers, the classification and naming of the shells gradually seemed to have become a generic and specific jungle. Up to the present more than 100 genera and subgenera have been described for approximately 400 extant species of the Pectinidae and the Propeamussiidae, the validity of many of those names is not certain and open to questioning. It is obvious that much work is still to be done on the taxonomy and nomenclature.

The descriptions of the species given have been taken where possible from the actual shells, either from museum collections, my own collection or other private collections. Where specimens of a certain species were unobtainable for study, which generally applies to the rarer species, I relied upon the original or subsequent diagnoses, which are found in the publications mentioned in the bibliography. In view of the nature of this book, which is intended to be a general guide only and not a scientific monograph, emphasis is placed mainly on a compact description of the valves and not so much on the sculpture or construction of the hinge system and/or the auricles, unless essential for the identification.

I have purposely not included a chapter on the anatomy of these bivalves, as such information can be found in most general shellbooks. I hope that the reader will overlook the errors and omissions bound to occur. C.D. Sherborn, the author of the famous Index Animalium, stated in 1922 that '...Although much time has been expended in trying to secure the endless combinations and permutations of specific names, it is felt to be impossible for one human being to attain completeness in this direction by reason of the colossal amount of literature to be dealt with'. The same, Mutatis mutandis, I feel, applies to a large degree to the study of the Pectinacea. I sincerely wish however, that this publication will be of practical use to shell-collectors, museums and Pecten-lovers at large.

ACKNOWLEDGEMENTS

I hereby wish to express my gratitude to the following persons and institutions for their assistance during the preparation of this book:

Dr. H.E. Coomans (Zoological Museum Amsterdam)
Mr. A. De Martino (Florida, U.S.A.)
Mr. H.H. Dijkstra (Zoological Museum Amsterdam)
Mr. P.L. van Pel (Zoological Museum Amsterdam)
Dr. W.F. Ponder (Australian Museum, Sydney)
Dr. H.A. Rehder (National Museum of Natural History, Washington)
Mr. J. Stanisic (Queensland Museum, Brisbane)
Prof. B. Tursch (Brussels, Free University), and
Mrs. Thora Whitehead (Brisbane);

all for specimens and/or informations, for help in research or loan of specimens, books and other publications.

Mrs. Shirley Slack-Smith (Western Australian Museum, Perth) and Dr. R. Willan (Queensland University, Brisbane) for constructive criticism at the manuscript stage.

Mrs. Joy Colpay for her great efforts in typing my hieroglyphic draft and last but not least my fellow shell collector Kevin Lamprell for his enthousiastic advice and assistance.

GLOSSARY

aculeate — Having prickles or spines.
adductor muscles — Posterior muscles which internally close the valves of the shell.
albino — White due to lack of colour pigments.
anterior — Front end, the opposite of posterior and the most remote end from the ligament.
auricle — Ear or wing.
auriculate With earshaped appendage.
axial — Applies to longitudinal ribs, bands or ridges.
beak — Extremity of bivalve from where the mature shell has developed or develops.
bifid — Divided by a groove in two parts, forked.
bivalve — Shell consisting of two parts connected by means of a hinge-system.
blotch — Irregular dot of contrasting colour.
byssal notch — Opening through which byssus protrudes.
byssus — Fibrous bundle by which bivalves attach themselves to rocks or other objects.
cancellated — Longitudinal ribs crossed by concentrical ribs of the same development, lattice-like — See also reticulated.
cardinal teeth — Interlocking teeth forming part of the hinge of a bivalve.
cartilage — Elastic material controlling the opening-movements of the valves.
cline — A geographically and gradually developing difference in appearance of certain populations of the same species.
concentric sculpture — Parallel running curved riblets or ridges but perpendicular to longitudinal or radial ribs.
concave — Curved or arched inwards.
conspecific — Belonging to the same species.
convex — Curved or arched outwards.
cord — Rounded lineal sculpture on shell-surface but smaller than ribs.
costate — Ribbed.
crenulated — Finely corrugated or notched.
dorsal — Back edge of shell towards outer hinge area.
ears — Flattened triangular shaped extensions of the valves near umbones of Pectinidae and Propeamussiidae. See also auricle.
equilateral — Having equal sides.
equivalve — When the two valves are of approximately equal size and form.
extant — Not extinct.
gape — Opening left between valves when shell is closed.
genotype — See type species.
genus — Group of closely related species sharing common origin and or structure.
growth lines — Lines marking boundaries of shell at different growth-stages.
holotype — Original specimen selected and used by an author to describe new species.
hinge — System of ligament and teeth to lock and unlock valves.
hyaline — Glasslike, transparent.
imbricate(d) — Arranged like tiles on a roof.
inequivalve — When valves are not of approx. same size and shape.
inflated — Swollen.
intercostate — Between ribs or ridges.
interspaces — Spaces between ribs.
interstices — Channels between ribs.
lamellae — Thin scales or plates.
lamellate — Covered with scales.

lamellose — With thin scales or plates.
left valve — The upper valve, sometimes concave, or flat to convex, more coloured, and less convex than the right valve, and lacking the ctenolium.
lirate — With fine raised lines.
longitudinal — Direction of ribs or lines from hinge to ventral edge. See also radiating.
maculated — Spotted or blotched, also called maculose.
margin — Edge or border of bivalve.
nodose — Having knobs or nodes.
oblique — Slanting, at an angle.
orbicular — With round shape.
pallial line — Perimeter of muscle-impression or scar inside bivalve.
pallium — Mantle.
pecten — Latin word for comb.
pelagic — Living in open sea.
plicate — With folds or twists, thereby forming ribs.
posterior — Rear or siphonal end of bivalve, opposite of anterior.
punctate — Dotted or spotted.
radial — Pertaining to a ray or to radiating.
radiating — Pertaining to ribs, striae or colours originating from central point or umbo at the top of bivalve and fanning out to ventral margin.
resilium — Internal cartilage in hinge-system.
reticulated — Network formed by cross-ridges. See also under cancellated.
rib — Radial elevation of bivalve surface.
right valve — The lower valve is mostly more convex than the left valve, less coloured or white, with a byssal notch and nearly always auricular denticles (ctenolium) near the byssal notch.
scabrous — Rough.
serrated — Saw-toothed.
species — Further subdivision of a genus of a group of shells having developed with common genetic properties by reason of environment and interbreeding.
spinose — With spiny or thorn-like sculpture.
striae — Rather flat or raised fine lines on valve surface.
subspecies — A group of local populations in a geographical subregion of the species, and taxonomically separated from other such populations .
synonym — A different name for the same species.
taxonomy — Study of rules and principles of classification.
type — The specimen on which the description of a (sub)species is based.
type species — Species selected by an author having general characteristics applying to certain genus.
umbo (plural umbones) — See beak.
undulated — Waved.
valve — One of the two detachable solid parts of a bivalve.
variant — A shell differing slightly in form or colour from other shells of a species.
variegated — With patches of different colours.
variety — See variant.
ventral margin — Outside edge of bivalve furthest away from hinge.
ventricose — Swollen.

Note: The following categories are used for sizes:
small - up to 25 mm long
medium - 25-60 mm long
large - 60 mm and over

TAXONOMY

Phylum	MOLLUSCA
Class	BIVALVIA
Order	OSTREOIDA
Superfamily	PECTINOIDEA
Family	PECTINIDAE
Family	PROPEAMUSSIIDAE

Aequipecten opercularis

Family PECTINIDAE

ADAMUSSIUM Thiele, 1935

Adamussium colbecki (Smith, 1902)
pl. 2 fig. 1

Orbicular, inequivalve, inequilateral, semitranslucent, thin, gaping, only slightly convex, medium size, with minute circular grooves and striae. Some have more pronounced ribbing than others. Small equal ears. Colour from white to purple. Inside iridescent. Type species of the genus, *from deep water Antarctica.*

AEQUIPECTEN Fischer, 1886

Type species *Ostrea opercularis* Linnaeus, 1758.
Shell solid, medium size, often rounded outline, equilateral, wider than high, left valve more convex than right valve. Both valves gaping, with from 18 to 24 fine, often irregular, radial ribs. No strong striae. Concentric growth lines visible in interspaces. Ears almost equal.

Aequipecten audouinii (Payraudeau, 1826)
pl. 1 fig. 1

Very similar in shape to *A. opercularis* and sometimes considered to be a subspecies thereof. Shell carries approx. 20 rather sharp, rough-feeling ribs. Unequal ears. Cream coloured background with orange-brown or lavender mottling. Right valve paler than left. Inside white with coloured edges. *Found in the Mediterranean Sea and adjacent Atlantic.* In the literature some subspecies are mentioned: *A.a. lamellosus* (Bucquoy, Dautzenberg & Dollfus, 1889), which has more lamellae, as the name indicates, and *A.a. transversus* (Clement, 1879), a broader specimen.

Aequipecten glyptus (Verrill, 1882)
pl. 1 fig. 3

Syn. *A. tryoni* (Dall, 1889)

Small to medium size, rather flat shell. Approx. 17 ribs which broaden and flatten out near the ventral edges. Concentric sculpture. Small equal ears. Colour one valve cream with pink rays on the top of the ribs, sometimes reddish-brown. Other valve dirty white, somewhat mottled with pink of brown. Inside also white. *Deep water from Cape Cod to Florida and Texas, USA*

Aequipecten heliacus (Dall, 1925)

Small to medium shell, equivalve, subcircular, slightly oblique. Approx. 20 narrow ribs with minute pointed scales. Near equal ears. Colour grey-white with some radial orange stripes, concentric mottling and blotches in pale lilac or orange. *West Indies to Florida.*

Aequipecten lineolaris (Lamarck, 1819)
pl. 1 fig. 7

Syn. *A. mayaguezensis* (Dall & Simpson, 1901)

Small to medium, semi-glossy, convex shell. Carries approx. 18 low folds, which inside the shell show up as ribbing, sometimes however more distinct outside ribs. Brick-red colour with many concentric fine and delicate lines of lighter shades, sometimes slightly mottled. Lower valve white. Equal ears. *Caribbean and S. Florida.*

Aequipecten minutus (Nickles, 1955)
pl. 23 fig. 1

Small, thin, flattish, somewhat elongated shell. Right valve more convex than left one. Both valves carry 20-21 rounded, radial riblets and numerous fine concentric striae. Ears unequal, colour dull reddish brown with some irregular whitish speckles, interior same colour but glossy. *West Africa from Senegal to Angola.*

Aequipecten muscosus (Wood, 1828)
pl. 1 fig. 5, 5a

Syn. *A. exasperatus* (Sowerby, 1842)
A. fuscopurpureus (Conrad, 1849)

Small to medium size. Both valves rather swollen and carry approx. 18-20 ribs with pronounced scales in their centre and with adjacent row of small lamellae. Touch, therefore, is rough. As an exception, however, some have smooth ribs. Colours range from lemon, orange, red to purple, often mottled with cream. Rather large subequal scrabrous auricles. *Carolina to Florida, Caribbean, Brazil.*
A. muscosus forma *acanthodes* (Dall, 1925), sometimes considered a distinct species, is a form, slightly less convex with more but shorter scales.

Aequipecten opercularis (Linneaus, 1758)
pl. 1 fig. 2, 2a, 2b

Syn. *A. subrufus* (Pennant, 1767)
A. pictus (Da Costa, 1778)
A. solaris (Born, 1778)
A. daucus (Reeve, 1853)

Medium to large size, thin, but strong, orbicular, almost equivalve, slightly oblique, convex shell. Valves carry approx. 20 radial, roundish, regular ribs, about equal in width with the channels in between. Fine riblets in these grooves. Valves often covered partly with tiny radially arranged prickles, which some times completely disappear with age. Margins serrated. Ears about equal. Colours very variable, either unicoloured, from whitish, yellow, pink, orange, brown, grey, purple-red, or with some cream or lighter or darker blotches or streaks of the basic colours. Right valve lighter in colour than the left one. Inside white with brownish or violet tinge. *East Atlantic from Norway to Cape Verde, Azores, North Sea and Mediterranean.*

Note. If one of the Pectinids has many synonyms, subspecies or varieties, it is certainly this European species. From more than 30 names I quote the following:

A. o. forma *albida* (Locard, 1888) white valves.

A. o. forma *aspera* (Bucquoy, Dautzenberg & Dollfus, 1889).

A. o. forma *bicolor* (Locard, 1888) two different colours.

A. o. forma *compressa* (Locard, 1888).
A. o. forma *concolor* (Bucquoy, Dautzenberg & Dollfus, 1889) uniform colour.
A. o. forma *depressa* (Locard, 1888) synonym of *A. o. planata.*
A. o. forma *inflata* (Locard, 1888) synonym of *A. o. tumida.*
A. o. forma *lateradiata* (Monterosato, 1884).
A. o. forma *lineata* (Da Costa, 1778) reddish concentric bands.
A. o. forma *marmorea* (Locard, 1888) marbled markings.
A. o. forma *monterosatoi* (Dautzenberg, 1889) small and smooth.
A. o. forma *perdix* (Bavay, 1905) has 26 ribs instead of the usual 20.
A. o. forma *planata* (Monterosato, 1884) very thin and flat valves.
A. o. forma *sanguineus* (Poli, 1795) synonym of *A. o. aspera.*
A. o. forma *squamosus* (Locard, 1888) more scaly than normal.
A. o. forma *tumida* (Jeffreys, 1863) more convex.
A. o. forma *undulatus* (Locard, 1888) with folds rather than ribs.
A. o. forma *vescoi* (Bavay, 1902) is a juvenile of *A. o. opercularis.*
A. o. forma *viridis* Lucas, 1981. This shell, *found in Brittany, France,* has some strong green, slightly yellowish blotches together with some deep red dots on the left valve. Right valve paler.

Aequipecten palmeri (Dall, 1897)
pl. 1 fig. 6

Small to medium size, slightly convex and equivalve. Approx. 14-15 narrow ribs, interstices with usually one central fine thread. Particularly the right valve carries rows of fine tiny scales making the shell feel rough. Large, wide, about equal ears. Cream or flesh coloured, often speckled, zigzagged or concentrically blotched with brown, or darkred. Right valve the darker. *Deep water Gulf of California, Mexico.*

Aequipecten phrygius (Dall, 1886)
pl. 1 fig. 4

Small, orbicular, equivalve, slightly convex shell with about 16-17 angular ribs with rows of fine lamellae. Deep interstices. Ears about equal. Colour grey blotched with pale rose rays. White inside with the rose shining through. *West Indies, Florida north to Cape Cod, USA..A. heliacus* is close to this species.

Aequipecten tamsi (Bavay, 1905)

Small to medium size, lightweight, almost equilateral and equivalve, slightly convex, right valve more so than left one. Slightly gaping near the ears which are unequal. Valves carry some 20 narrow slightly angular ribs with minute scales. Colour outside greyish-brown with small grey spots. Inside pale-brown. *Venezuela.*
Close to *A. opercularis.*

AMUSIUM Röding, 1798

Type species *Ostrea pleuronectes* Linnaeus, 1758. Delicate, thin, equivalve, orbicular shell from small to large in size. Both valves gaping and convex. Shell smooth and usually glossy outside, but often with interior radial ribbing. Ears mostly equal, moderate to small in size. Right valve has darker colours than left valve and is sometimes blotched.

Amusium balloti (Bernardi, 1861)
pl. 2 fig. 5, 5a

Large, but as a rule somewhat smaller than *A. japonicum,* thin, smooth, orbicular, glossy,

with gaping sides and slightly convex. Left valve orange-brown to brown or purple, often with numerous concentric lines or bands and or radial segments in darker or lighter shades, inside white. Right valve white, also inside white. Small equal ears, which are bluish rose. Inside each valve 42-50 small ribs. Glossy white background with a touch of lavender in the umbonal areas. *S. Pacific including Australia.*
Sometimes considered a subspecies of *A. japonicum.*

Amusium japonicum (Gmelin, 1791)
pl. 2 fig. 2

Large, circular, thin shell with small equal ears. Both valves gaping. Convex, left valve more so. Smooth and glossy on outside, ribbed inside, totalling from 48 to 54 on the right valve which is brick-red. Sometimes with numerous brownish concentric lines of various width. Ears also red. Left valve creamy white, also ribbed on the interior side which is glossy and white usually with bright yellow rim along the margins. *Japan and W. Pacific.*

Amusium japonicum formosum Habe, 1964 (non Melvill & Standen, 1907)

This subspecies has the same size and internal ribformation as *A. japonicum* ss. Number of ribs on right valve from 46-50, on left valve 35-44. Exterior of left valve shows numerous brown lines along the concentric growthmarks. Yellow margins on the inside. Ears are deep brown. *Japan, Taiwan, China .*

Amusium laurentii (Gmelin, 1791)
pl. 2 fig. 4

Medium to large in size, smooth and glossy. The less convex valve reddish brown with ray patterns or mottled. The other valve, more convex, is much lighter in colour and often off-white. The flatter valve, being smaller, rests in the convex one. Insides of the valves have approx. 20 sets of 2 regular, rounded ribs. Colour interior cream. About equal ears. Shallow to deep water. *Carribean.*

Amusium obliteratum (Linnaeus, 1758)
pl. 2 fig. 3

Medium size, light, orbicular compressed, smooth, glossy shell, valves gaping and serrated at the margins. Numerous internal ribs, visible through the translucent valve, but no external ribbing. Equal ears. Colour upper valve red-brown, purplish near margins, sometimes radial stripes in lighter or darker shades. Lower valve considerably lighter in colour. *West Pacific, described from China Sea.*

Amusium papyraceum (Gabb, 1873)
pl. 3 fig. 1

Small to medium size, shell flat to moderately convex. Smooth and glossy. Somewhat gaping. No exterior ribbing, interior with many fine riblets in pairs. Upper valve red purple to reddish brown, with some darker blotches or darker concentric lines or bands. Lower valve white, sometimes with yellowish margins. Rounded hinge line. *West Indies, from the Gulf of Mexico to Brazil, in deep water.*

Amusium pleuronectes (Linnaeus, 1758)
pl. 3 fig. 3

Syn. *Pleuronectia laevigata* Swainson, 1840

Rather large but smaller than *A. japonicum.* Almost circular, often glossy, smooth, flat shell, gaping at the sides. Has approx. 13 pairs of slightly elevated ribs on the inside of the valves. Small equal ears. Colour left valve varies considerably from grey, pink, purplish to deep brown, with radiating bluish bands or sections and white spots between the radial bands. Right valve white, interior white. *West Pacific from Japan to Australia and the Indian Ocean.*

Amusium pleuronectes australiae Habe, 1964

This subspecies has the same size, but left valve is dark pink with bluish markings or rays which usually become narrower towards the margins and white dots in the umbonal area. Right valve white. Inside white. Shell has less ribs than *A. pleuronectes* s.s. *Northern parts of Australia.*

ARGOPECTEN Monterosato, 1889

Type species *Pecten solidulus* Reeve, 1853.

In the past considered a subgenus of *Aequipecten* and of *Chlamys.* Medium sized shell, thick, average to extremely convex, round in shape with numerous strong, mostly angular ribs, not spinose. Valves shiny, not gaping. Ears unequal. Species occur in abundant colour variations, cloudy patches or spots usually in purple or brown hues.

Argopecten asperulatus (Adams & Reeve, 1850)
pl. 23 fig. 2

Small, irregularly multi-ribbed, equivalve somewhat oblong scabrous shell. Ears very unequal. Brownish or orange in colour with cream mottling, particularly from the umbo areas down to mid-body. *Korea, Japan, Philippines, Indonesia in shallow water.*
A. pelseneeri Dautzenberg & Bavay, 1912 is sometimes considered a subspecies of *A. asperulatus.*

Argopecten circularis (Sowerby I, 1835)
pl. 3 fig. 4, 4a

Syn. *A. tumidus* (Sowerby I, 1835)
A. inca (D'Orbigny, 1846)
A. ventricosus (Sowerby II, 1842)
A. solidulus (Reeve, 1853)
A. filitextus (Li, 1930)

Solid medium to large size shell, somewhat oblique, very inflated, equal valves and equal ears, with 17-22 smooth and wide ribs, which flatten towards the ventral edges. Colours and pattern very variable, from off-white to uniform orange, brown or violet, often with concentrically arranged but irregular blotches or streaks. Left valve darker than right one. Usually has internally a wide chocolate-brown band on whitish background, varying in width and intensity. *California to Peru.*
Somewhat similar in appearance to its twinspecies from the Gulf of Mexico, *A. gibbus* (Linnaeus, 1758).

Argopecten circularis aequisulcatus (Carpenter, 1864)

Differs from the nominate subspecies in being less convex, thinner, with narrower ribs and softer colouring, often speckless. *Northern California*

Argopecten commutatus (Monterosato, 1875)
pl. 3 fig. 2

Syn. *A. philippii* (Recluz, 1853)
A. *atlanticus* (Smith, 1890)

Small to medium, circular, equivalve, strong shell, moderately globose, with approx. 20 flat rounded radial ribs with deep interstices. Further radial sculpture and small concentric lamellae. Ears unequal. Narrow byssal notch. Colours off-white with red or brown stripes, blotches or wavy patterns. Sometimes uniformly violet with deeper purple blotches or even yellow. Inside white or soft violet. *Atlantic Ocean from France to Canary Islands and Azores, West Coast of N. Africa and Mediterranean.*
Locard (1888) described several varieties: *depressus, globulatus, major, minor* and *transversus.*

Argopecten flabellum (Gmelin, 1791)
pl. 3 fig. 5, 5a

Small to medium smooth shell, equivalve, very convex. Approx. 20 regular ribs on both valves. Near equal ears. Colours vary from cream to orange, pink purple with brown and white blotches, sometimes uniformly coloured. Interior white, margins have same colour as the one dominating the exterior. *West Africa.*
This shell seems to be the twin species of *A. gibbus* from the West Atlantic USA.

Argopecten gibbus (Linnaeus, 1758)
pl. 3 fig. 6, 6a

Syn. *A. liocymatus* (Dall, 1925)
A. purpuratus (Conrad, 1831 non Lamarck, 1819)

Small to medium size, strong, near equilateral, equivalve, convex shell. Shell carries 17-25 ribs, becoming broader towards the ventral edges. Sometimes with radial ridges and grooves. Some concentric lines showing in interstices, the latter about equal in width to the ribs. Almost equal ears. Very colourful. Left valve from cream to violet with orange, pink, red, brown or violet waving or mottling. Often however uniformly coloured. Right valve white or cream with paler additional colours. Interior white sometimes bordered in pink. *Gulf of Mexico, W. Atlantic, Florida to Delaware Bay USA.*
A. gibbus forma *portusregii* (Grau, 1952) replaces the name *A. gibbus carolinensis* (Grau, 1952) (preoccupied). The valves are somewhat flatter and have pronounced concentric lamellae in interstices. *Off Port Royal, South Carolina.*
A. gibbus forma *sowerbyi* (Guilding, 1826) is supposed to be a colour form from the *Antilles.*

Argopecten irradians (Lamarck, 1819)
pl. 4 fig. 1, 1a

Syn. *A. borealis* (Say, 1822)
A. turgidus (Gmelin, 1791)

Strong, circular, medium to large shell, equivalve and equilateral, rather convex, somewhat gaping, with 14-20 radial, rounded ribs, undulated growthlines showing across. Interstices narrower than ribwidths. Ears about equal in size. Colour: upper valve very variable from white, yellow, orange, brown, greyish and even close to black. Lower valve, the more convex one always lighter in colour or white. Colour uniform, radiating or mottled. Inside glossy white. *East coast USA from Cape Cod to New Jersey.*

Argopecten irradians concentricus (Say, 1822)
pl. 4 fig. 2

Medium size, equilateral, valves gaping anteriorly, more convex and thicker than the nominate subspecies. Right valve usually off-white, left valve greyish to brown or darkbrown sometimes whith indistinct mottled or banded design. *Maryland to Georgia, Tampa (Florida) to Louisiana.*

Argopecten irradians amplicostatus (Dall, 1898)

Smaller in size, convex, valves gaping slightly or not at all. Valves carry fewer ribs than *A. i. irradians* (approx. 13-17). They are high and rounded and wider than interstices. Concentric lamellae showing in these interspaces. This subspecies has thicker and more convex valves than *A. i. concentricus.* Colour same as the latter. *Texas to Mexico, and Colombia.*

Argopecten noronhensis (E.A. Smith, 1885)

Small, slightly inequivalve, oblique rather convex shell. Right valve has 17 smooth broad ribs, interspaces narrower and sculptured with fine concentric lamellae. The ribs, also 17 in number on the left valve narrower and the grooves are equally wide. The costae are divided in two parts by a central lira and crossed by subtile lamellae. Unequal ears. Colour white, blotched with deep rose or orange. Internally purple with white margins. Smaller specimens seem to have no distinct internal colour. *Described from West Atlantic near Fernando Noronha, Brazil.*
Not unlike *A. gibbus,* but more convex, ribs shaper.

Argopecten nucleus (Born, 1778)
pl. 4 fig. 3

Shell small to medium sized, convex, equilateral and approx. equivalve. Valves carry 18-22 ribs which are wider than the interstices, where some concentric lamellae are present. *S. E. Florida, Bahamas, West Indies, Gulf of Mexico.*

This species is different from *A. gibbus* in that it has overall a more whitish colour on the left valve, lesser defined valve flanks and generally more convex.

Argopecten pelseneeri (Dautzenberg & Bavay, 1912)
pl. 6 fig. 3

Syn. *Pecten rugosus* Sowerby II, 1842 (non Lamarck, 1819, a fossil)

A small to medium species, with approx. 20 slightly rough, radial ribs, brown to flesh-coloured, often with whitish streaks, blotches or speckles. Inside cream with soft yellow, brown, or purple. Very unequal ears. *West Pacific from Thailand to Japan.*

Argopecten purpuratus (Lamarck, 1819)
pl. 4 fig. 5

Large, solid, circular, moderately convex, equilateral, equivalve shell. Valves carry 22-24 broad, squarish, smooth ribs, interstices lamellated. Left valve lamellated on ribs as well. About equal ears. Large byssal groove. Colour usually white with purple on top of the ribs, alternately pink, brown. Mottling no exception, sometimes all white, cream or orange with cream and purple. Inside of white shells white, coloured specimens have corresponding concentric colour bands inside. *Central America to Chile.*

Argopecten rehderi (Grau, 1960)

Small shell, subequivalve, well rounded, only moderately convex. Right valve has 18-23 rounded ribs centrally arranged and 4 or 5 riblets flanking each submargin, with fine concentric lamellae usually worn off at the top of the ribs. Left valve slightly deeper than right, with identical ribbing and sculpture. Colour right valve white cream or pink speckled with brown, often wavy streaks of white on ribs. Left valve stronger coloured streaks or blotches of brown or red-brown. *French Polynesia.*

Argopecten schrammi (P. Fischer, 1860)

Medium size, broad, semi-orbicular, nearly equivalve, somewhat depressed. Valves bear about 20 rounded radial ribs. Interspaces equal to the width of the ribs. Showing some concentric lines. Colour greyish with brown mottling. Inside brown. Near equal ears. *Caribbean.*

BRACTECHLAMYS Iredale, 1939

Bractechlamys evecta Iredale, 1939
pl. 28 fig. 1

Small to medium size, oblong, only slightly convex shell, right valve somewhat more than the left one. Both valves have approx. 9 fold-like ribs and are completely covered with fine radial, scaly, cords or striae. Ears very unequal. Colour orange to brick-red or even deep purple. Inside white with shade of the same colour as outside, but very glossy. *Coastal waters Queensland, Australia.*
Type species of *Bractechlamys*. By some authors considered a synonym of *Comptopallium vexillum*.

CHLAMYS Röding in Bolten, 1798

Type species: *Pecten islandicus* Müller, 1776.

Oval-oblong small to large shell, slightly oblique with moderately convex and almost equal valves which are corrugated by 17 or more strong ribs, often supported by riblets in the interspaces. Ribs and riblets generally having scales or spines; some shells however might be called smooth. Internal surface corrugated, following the external ribbing, ventral margin waved. Ears, often lamellated with scrabrous edges, large but unequal in area, anterior one the larger. Large byssal notch. Great range of colours in the genus.

Note. With regard to the multitude of subgenera of this genus it might be useful to point out that most of these are only based on one or a few of the characteristics of the type species. Even the type species sometimes shows variations or deviations which do not always possess the properties deemed to be diagnostic of the genus as a whole.

subgenus CHLAMYS

Chlamys (Chlamys) albolineata (Sowerby II, 1842)
pl. 6 fig. 4

Small to medium size, compressed shell, slightly oblique. Unequal ears. Numerous small squarish, rough-feeling ribs, deep intercoastal areas. Cream coloured, grey spots on lower valve, upper valve grey, yellow or orange or brownspeckled or streaked against a white background. Some riblets remain white. Inside glossy white. *Japan to S. W. Pacific and S. Pacific.*
Very close to *C. ornata* but has more ribs. Also close to *C. irregularis* and sometimes considered a synonym hereof.

Chlamys (Chlamys) alta (Reeve, 1853)

Elongated, rather convex, equilateral and equivalve shell. Valves covered with numerous smooth and thin ribs. Ears unequal. Colour lavender, here and there blotched with dark brown. *West Pacific.*

Chlamys (Chlamys) argentinensis Castellanos, 1973

Small, thin, fragile, somewhat orbicular, slightly convex shell. Valves carry 18 radial main ribs, each consisting of 3 smaller ribs, which are getting more prominent from the centre of the valves towards the margins. Intercostal grooves have a square profile and are covered with numerous fine concentric striae. Ears unequal. Colour of the shell bright white with pink or yellowish centres. *Gulf of San Matias, Southern Argentina.*

Chlamys (Chlamys) benedicti Verrill & Bush, 1897
pl. 23 fig. 4

Tiny, fragile shell, with 22 distinct ribs and minute ones in between. Also a number of regular elevated concentric lines resulting in scaly or spiny points at the crossings. Unequal ears. Colour range from yellow, orange, pink, red, through brown to a light purple, usually with white zigzags. *From Caribbean waters to Texas in shallow depths.*
See also under *C. nympha.*

Chlamys (Chlamys) bruei (Payraudeau, 1826)
pl. 23 fig. 5

Syn. *P. sulcatus* Jeffreys, 1879.

Small, fragile, not very convex, equivalve. Left valve carries approx. 20 irregular ribs with thin scaly cords on top. Small spines also in interstices. Right valve has from 15-20 groups of two rounded scaly ribs. Often concentric growth lines noticeable. Ears very unequal. Colour shades of off-white, yellow, orange to brownish-purple. *Atlantic to Cape Verde Islands and Mediterranean.*
C. bruei forma *pulchricostata* Nordsieck, 1969 is larger with 24 ribs, colour orange-brown.
C. bruei forma *coeni* Nordsieck, 1969 is elongated, and also larger than the typical, colour purple with some white mottling. Often white spot at umbones. *Adriatic Sea.*

Note. These two formae were described as subspecies by Nordsieck.

Chlamys (Chlamys) chosenica Kuroda, 1932

Small to medium, slightly convex shell with unequal ears, about 24 radial ribs. Flesh-coloured to violet. Closely resembles *C. hindsii* but differs in having a more elongated shell with more regularly arranged ribbing. *Korea, Japan.*

Chlamys (Chlamys) cloacata (Reeve, 1853)
pl. 4 fig. 4

Small to medium size, equivalve, convex. Valves bear 16 rounded and raised, rough ribs with deep interstices, rather widely spaced. Unequal ears. Colour basically whitish with stains in pale brown and flesh colours. *West Pacific.*

Chlamys (Chlamys) coruscans (Hinds, 1845)
pl. 23 fig. 7, 7a, 7b

Syn. *P. cuneolus* Reeve, 1853
P. sulphureus Dunker in Küster & Kobelt, 1888

Small, thin, slightly elongated, moderately convex, with about 15-16 radial, rounded, ribs or folds, grooved by 4 threads which are also present in the intercostal areas. Unequal ears. Colour varies, usually yellowish cream, variegated with yellow, brown or wine-red, with some small white streaks near umbones. In some cases uniformely orange. *Indo Pacific.*

Chlamys coruscans hawaiensis (Dall, Bartsch & Rehder, 1938)

Much the same as the above. Difference mainly in the ears. In this subspecies the posterior ear is almost non-existing. *Endemic to the Hawaiian Islands.*

Chlamys (Chlamys) cristularis (Adams & Reeve, 1850)
pl. 5 fig. 1

Syn. *P. indicus* Deshayes, 1832

Small orbicular, solid shell, somewhat elongated and oblique, convex. Bears about 25 flat prickly ribs. Ears relatively large, with scales on the top. White background with red mottling. *Indo-Pacific.*

Chlamys (Chlamys) crotilus (Reeve), 1853

Tiny shell, typical *Chlamys* shape, valves only slightly convex with numerous scaly riblets. Ears very unequal. Colour purple-red with darker blotches. Locality not mentioned by Reeve.

Note. As this small shell never turns up in collections, I suspect the description is of a juvenile *Chlamys* species.

Chlamys (Chlamys) cytherea Bavay, 1904

Small to medium, strong, rounded, slightly convex, equilateral but inequivalve species. Both valves, of which the upper valve is the more convex, carry 16 grooved ribs, here and there slightly scabrous. Interspaces narrow, decorated with one or two radial rows of fine scales. Large subequal, triangular ears. Colour pinkish-white purple mottling outside, inside a light purple.

Note. Described from only one specimen. Similar to *Aequipecten muscosus.* Since the latter shell originates from the *Caribbean*, I suspect that "Taiti" in the original description should read *Haiti.*

Chlamys (Chlamys) denticulata (Adams & Reeve, 1850)

Small shell, elongated, rather flat, thin, translucent almost equivalve shell. Left valve carries about 15 radial irregularly shaped ribs with small scales. Right valve carries a very large number of scabrous riblets. Ears very unequal and denticulated at the top. Colour yellowish with dashes of soft rose. *Southern Pacific.*
Some authors consider this species synonymous with *C. hastata.*

Chlamys (Chlamys) dichroa (Suter, 1909)
pl. 5 fig. 2

Small to medium size, solid, fan-shaped, only slightly convex, valves with approx. 18 radial ribs and deep intercostal spaces. Whole shell features close rough cords developing into scales on the ribbing. Ears unequal. Colour orange to brick-red background, variegated with cream or darker red blotches. Often obtained from the stomach of cod fish. *Stewart Island, New Zealand, deep water.*

Chlamys (Chlamys) dieffenbachi (Reeve, 1853)
pl. 5 fig. 4

Syn. *C. celator* Finlay, 1928

Medium size, lightweight shell, both valves considerably convex. Numerous thin ribs with riblets and striae in between. Short sharp scales. Very unequal ears. Colour varies from yellow, pink, orange or often greyish purple, sometimes with white rays. Inside same tints but stronger. Cream on ears and marginal areas. *New Zealand.*

Note. It has been suggested that this taxon is a larger form of *Chlamys zelandiae. C. suprasilis* Finlay, 1928, synonym *C. crepisculi* Fleming, 1948 is most likely a subspecies.

Chlamys (Chlamys) elegantissima (Deshayes, 1863)
pl. 24 fig. 1

Described from only one right valve, although Bavay in 1902 elaborated the description from some complete specimens. Valves carry about 25 unequal radial fine scaled ribs. Very unequal ears. Both valves white with rose mottling. *Holotype from Reunion, Indian Ocean.*

Chlamys (Chlamys) empressae Kuroda & Habe, 1971
pl. 23 fig. 8, 8a

Small shell, oblique, very unequal ears. The valves carry a large number of fine radial ribs with fine scales, particularly near the ventral edges. Soft lavender pink or orange background. Often with concentric bands or speckled with tiny white flecks. Inside as outside. *Japan, Okinawa.*

Chlamys (Chlamys) erubescens (Reeve, 1853)

Small to medium size, rounded shell. Near equal ears. Right valve convex, left valve considerably flatter. Both valves have 20 narrow somewhat rough ribs and deep interstices. Colour white, pink tinged. Right valve paler. Description taken from Reeve, who does not quote a locality.

Chlamys (Chlamys) farreri (Jones & Preston, 1904)
pl. 5 fig. 7

Syn. C. nipponensis Kuroda, 1932

Typical *Chlamys,* medium to large size, convex, thin, carrying about 10-16 very scaly main ribs and numerous secondary, slightly scaly, ribs and threads. Very unequal ears. Colour white, orange, brick-red or purple. Sometimes unicoloured, often with white occasionally with purple-brown on ventral margins and in umbonal areas. *Japan, Korea and adjacent waters.*

Chlamys farreri akazara Kuroda, 1932 is hard to distinguish from the typical. This shell seems to be more slender and deeper in colour, pale purple or reddish inside. *Japanese waters.*

Chlamys (Chlamys) felipponei (Dall, 1922)

Small to medium size, rather smooth shell, somewhat oblique, convex. Left valve has 5 well-spaced rounded main ribs and further very fine striae in between. Concentric growth lines, though almost invisible on the right valve. Unequal ears. Colours: left valve flamed with white, red or rose, usually with irregular bands. Right valve lighter. *Argentina.*

Chlamys (Chlamys) fricata (Reeve, 1853)

Small to medium size rather orbicular shaped, equivalve, flattish, but somewhat convex in central valve area. Large number of flat ribs (approx. 25), finely scaled. Unequal ears. Colour off-white with some reddish mottling, especially around the margin and on the ribbing. *W. Pacific, N.E. Australia, Andaman Islands.*

Chlamys (Chlamys) gilchristi Sowerby III, 1904
pl. 24 fig. 2

Small to medium size shell, circular, equivalve, thin and rather flat. Left valve carries 9 radial ribs, also numerous radial, granulose lirae. Right valve more convex, ribs broader sculptures with fine concentric lamellae. Unequal distinctly sculptured ears. Colour left valve orange-yellow, right valve pinkish hue. Inside a soft golden orange. *South Africa, Cape Province.*

Chlamys (Chlamys) hastata (Sowerby II, 1842)
pl. 5 fig. 8, 8a

Thin, rough, somewhat elongated, medium size, only slightly convex shell with serrated margins. Right valve bears approx. 20 main ribs with distinct fine spines and further smaller riblets and smaller spines. Left valve has approx. 10 scaled ribs with a further 12-15 beaded ribs or cords and numerous radial threads in the interspaces. The spines or scales are rather curly. Unequal ears. Very colourful in bright lemon, orange or red. Left valve darker. *California.*

Chlamys hastata hericius (Gould, 1850)

Medium size, thin. There is some difference in the ribbing. Spines are shorter. Colour varies: white, cream, yellow, pink, rose and combinations thereof, often in ray-fashion. *Shallow water, Alaska to California.*

Chlamys hastata pugetensis (Oldroyd, 1920)

This subspecies is different from the above by having less ribs. Is somewhat smaller and its interior colour is whitish, very glossy with brown or red in the margin areas and minute sculpture not present in the others. *Alaska to California.*

Chlamys (Chlamys) hindsii (Carpenter, 1864)
pl. 5 fig. 5

Medium size shell. Right valve rather flat with smooth and rounded ribs, often in pairs. Left valve has numerous thin ribs which each carry 3 rows of small spines. Unequal ears, distinct byssal notch. Colour: pale hues of cream, yellow, orange, rose or mauve. *Alaska to Mid-California.*
Sometimes considered a synonym of *Chlamys rubida* (Hinds, 1845).

Chlamys hindsii navarchus (Dall, 1898)

Very much like the typical, but ribs are smoother and more even. Lavender colour. Grant & Gale (1931) placed it in the synonymy of *Chlamys hindsii* s.s.

Chlamys (Chlamys) humilis Sowerby III, 1904
pl. 24 fig. 3

Small, thin, elongated, equilateral shell. Valves carry about 32 small slightly imbricated but distinct, squamose radial ribs, and a further 15 minor radial riblets in between. Right valve rather convex, left valve nearly flat. Ears very unequal. Colour: reddish orange with some yellowish rays. *South Africa.*

Note: Barnard (1964) stated that he can hardly distinguish this shell from *Chlamys tincta*. Another related species is *C. weberi*. According to Wagner (1983) one has to look at the microsculpture to distinguish the 3 species. In *C. tincta* this is made up from small grooves between the ribbing. *C. weberi* shows reticulation between the ribs, whereas *C. humilis* has grooves and concentric microlamellae on the ribs' edging.

Chlamys (Chlamys) incantata, Hertlein, 1972
pl. 6 fig. 9

Medium size, oval shaped shell, about equivalve. Valves decorated with approx. 25 low and narrow ribs, each carrying a row of hollow spines. Flat intercostal spaces with some fine radial spiny cords, also fine concentric growth lines. Ears unequal with large byssal notch. Colour off-white with rosy spines. *Galapagos Islands.*

Chlamys (Chlamys) islandica (Müller, 1776)
pl. 28 fig. 3, 3a

Syn. *Ostrea cinnebarina* Born, 1778
P. paleii Conrad, 1831

P. fabricii Philippi, 1844 (juvenile)
C. insculpta Verrill, 1897
P. costellata Verrill, 1897 (juvenile)
C. scotica Simpson, 1910

Large shell, thin, elongated, equivalve, nearly equilateral, convex. Valves bear approx. 50-60 rather irregular low, thin, radial ribs, often in groups of 2 or 3, coarse but not spiny or scaly, varying in width. Growth lines showing. Ears rather unequal. Colours from streaky off-white, peach, orange-red to purple, also concentric bands in darker or lighter shades. Left valve usually much lighter in colour. Inside somewhat as outside but in softer colours. *Deep water over large areas from the Arctic down to the Northern regions of the Amphiatlantic and Pacific waters.*

Chlamys islandica beringiana (Middendorff, 1849)

Syn. *P. strategus* Dall, 1898

Differs from the nominal subspecies mainly in the structure of the ribs which are coarser and broader. Also shows some cross-hatched sculpture. Inside glossy white. *Bering Sea, Alaska, British Columbia.*

Chlamys islandica erythrocomata (Dall, 1907)
pl. 4 fig. 6

Generally as *C. islandica* s.s. but colour off-white to soft salmon, banded and differs further by having broader ribs. Umbonal areas pink to purplish in colour.

Chlamys islandica forma *albida* Dall, 1906

Is a light colour form, white to pale pink. *Bering Sea to Japan. Deep water.*

Note. To quote Dr. H. Rehder who examined the holotypes of this subspecies and of *C. islandica erythrocomata:* 'I would be inclined to unite them' (cf. Grau 1959:73).

Chlamys islandica wainwrightensis MacNeil, 1967

This is another subspecies or form of *C. islandica.* Ribs are coarser and broader than of the latter. *Alaska.*

Chlamys (Chlamys) jousseaumei Bavay, 1904
pl. 6 fig. 5

Syn. *C. parirregularis* Nomura & Hatai, 1940

Shell small to medium, thin, oblique, rather flat. Valves show up to 30 fine scabrous irregular riblets with relatively large interspaces, which in turn show fine radial striae. Most of the valve area traversed by fine concentric grooves, unequal ears. Colour pinkish, the riblets being coloured in a darker brick-red with some cream mottling or flames. Outer colours visible through the inside. *Japan and also reported from Indonesia.*

Note. See also under *C. mollita* and *C. princessae.*

Chlamys (Chlamys) kauaiensis Dall, Bartsch & Rehder, 1938

Very small shell, rounded, translucent with about 25 regular, narrow, radial riblets or threads, which are visible from the inside of the valves. Subequal ears. Colour white. *Hawaiian waters.*

Chlamys (Chlamys) kiwaensis Powell, 1933
pl. 6 fig. 7

Medium size, thin, suborbicular, oblique, convex shell, unequal ears. Valves carry sharp narrow radial ribs with scattered fine thorny sculpture. Interstices with fine threads. Colour of both valves creamy white, ribs on left valve brownish. Comparable with *C. delicatula. North Island, New Zealand. Very deep water.*

Chlamys (Chlamys) layardi (Reeve, 1853)

Small to medium size shell, oblong, convex. Carries approx. 20 sharp slightly spiny ribs. Ventral edge serrated. Colour deep brown with light mottling or cream with brown mottling. *Indo-Western Pacific.*

Chlamys (Chlamys) lemniscata (Reeve, 1853)
pl. 6 fig. 2

Small, slightly convex, equivalve, slightly elongated and oblique shell. Numerous scaly radial riblets and fine tiny scales in the interstices. Unequal ears. Cream colour with irregular red-brown to purple blotches sometimes forming an attractive concentric pattern. Inside glossy, white with the outside pattern of the right valve visible. *Indian Ocean, Australia, West Pacific.*

Note. By some authors considered a subspecies of *Chlamys (Scaeochlamys) irregularis.*

Chlamys (Chlamys) liltvedi Wagner, 1984

Medium size shell. Left valve, more convex than right valve, carries 9-13 pronounced main ribs and 22-26 finer riblets in the interspaces. Right valve has 22-32 pronounced ribs and 10-15 finer riblets in between. Unequal ears. Colour: orange, with darker and slighter shades of the same colour, yellow with brown, red in shades, or purple with brown, yellow and white near the umbones. *South Africa, Natal, particularly the coast of Zululand.*

Note. This is most probably the shell called *Chlamys squamosa* (Gmel.) by Barnard (1964). However *C. liltvedi* has a less swollen right valve and the ribs, distinctly pronounced and partly less pronounced, are more distinct from the apex downwards. In *C. squamosa* only faintly visible for the first quarter of the total height of the valve. The latter has a far wider distribution.

Chlamys (Chlamys) lowei (Hertlein, 1935)
pl. 24 fig. 4, 4a

Rather small elongated shell, valves almost equal, covered with approx. 20 sharp ribs with rows of tiny spinose knobs, ribs wider than the interspaces which have a coarse riblet. Ears are very unequal, the anterior one being the largest. Colour brownish in ventral area and mottled in cream, grey and brown, exceptionally orange or reddish brown. *Deep water California to Ecuador, Galapagos Islands.*

Chlamys (Chlamys) marshallensis Waller, 1972

Small, oblique shell with unequal ears. Left valve is slightly smaller and fits in larger right valve. Both valves carry approx. 30 scabrous riblets and are covered for most of the valve area by a pattern of fine diverging striae. Colour white with reddish rays or with radially elongated spots, sometimes in yellow, orange or red shades. *Marshall Islands, Pacific.*

Note. Resembles *C. squamosa.* Considered by some authors a synonym of *Chlamys wilhelminae* Bavay, 1904.

Chlamys (Chlamys) mildredae (Bayer, 1941)
pl. 6 fig. 6

Small, flattish, oblique shell. Both valves carry about 30 riblets, some smaller, some larger, with fine scales and striae on the ribs. Unequal ears. Exterior colours white, orange, red, brown or purple or mottled on the left valve. Right valve much paler or even white. Inside pale yellow with purple near the edges. *Gulf of Mexico, Florida, Bermuda.*

Note. Close to *C. sentis.*

Chlamys (Chlamys) mollita (Reeve, 1853)
pl. 6 fig. 8

Small to medium size, thin, equivalve, equilateral, only slightly convex, translucent shell with numerous irregular small striae. Very unequal ears. Colour reddish brown with cream or tan rays. Inside rose-coloured. *Japanese waters.*

Chlamys (Chlamys) multisquamata (Dunker, 1864)
pl. 6 fig. 10, 10a

Syn. *P. effluens* Dall, 1886

Medium large, flat, fragile, inequivalve shell. Valves have numerous fine, irregular radial cords, very smooth near apex. Small unequal ears. Colours lemon, cream or orange with brownish-pink spots, sometimes deep red markings. *Bahamas, Cuba, and Florida.*

Chlamys (Chlamys) multistriata (Poli, 1795)
pl. 24 fig. 5, 5a

Small to medium size, oblique, convex, both valves with approx. 50 rounded riblets. Feels rough because of the minute scales on the ribs, which sometimes also occur in a concentric fashion in the interstices. Ears unequal. Variable colours from off-white orange, reddish-brown to dark brown, some whitish irregular markings. *Deep water Atlantic to Canary Island, Mediterranean.*

Note. Many forms were described: *elongata* Locard; *flava, rubra* and *violacea* by Monterosato.

Chlamys (Chlamys) nivea (MacGillivray, 1825)
pl. 7 fig. 1

Medium size, light, strong, moderately convex, circular shell. Left valve the more convex one. Gaping slightly towards the umbo. Valves have over 40 thin finely scaled ribs. Very unequal ears. Colour creamy white, sometimes with orangy mottling, dull. Inside glossy white. *North Atlantic, particularly the West coast of Scotland.*

Chlamys (Chlamys) nympha (Bavay, 1906)

Small, thin, somewhat translucent, slightly oblique, inequivalve, lower valve the more convex one. Valves carry approx. 22 small triangular ribs with fine scales. Small unequal ears. Colour pinkish on the top valve with slightly lighter coloured rays, lower valve paler. Described from specimen ex *Caribbean.*

Note. Some authors consider this taxon synonymous with *C. benedicti* (Verrill & Bush, 1897)

Chlamys (Chlamys) ornata (Lamarck, 1819)
pl. 7 fig. 2, 2a

Small, equivalve, only slightly convex, oblique shell with approx. 18 scaly ribs on top valve alternated with 2 prickly striae. Ribs and scales often worn flat. On bottom valve riblets arranged in 18 groups of 3. Very unequal auricles. Colour off-white or cream spotted or blotched in purplish tones, somewhat in a concentric fashion. Left valve paler. Inside translucent white with blotches shining through. *Shallow water West Indies, Florida, Brazil.*

Chlamys (Chlamys) phalara Roth, 1975
pl. 7 fig. 3

Medium to large, rounded somewhat oblique, slightly convex, left valve more so than right one. Left valve has about 35 narrow single ribs from smooth to scaly, scales short and raised. Intercostal areas show thin concentric lamellae. Right valve approx. the same number of riblets, often in groups of 2 or 3, smoother at the umbo but frilly at the valve margin. Ears unequal, distinct byssal notch. Colour from white, cream to yellow or reddish, uniform or variegated, sometimes with V-shaped white streaks. Right valve paler. *Juan Fernandez Island, Chile, in deep water.*

Chlamys (Chlamys) princessae Kuroda & Habe, 1971
pl. 24 fig. 6

Small convex, light shell. Valves with numerous fine riblets or threads. Ears unequal. Flesh-coloured, speckled or mottled with cream. *Japan.*

Note. Like *C. jousseaumi*

Chlamys (Chlamys) pseudolima (Sowerby II, 1842)

Medium size shell, equivalve ovalround, slightly oblique, thin, convex. Valves carry approx. 27 rounded ribs each sculptured with 3 rows of close, sharp erect scales. Very unequal and distinctly scaly ears. Colour variegated in orange and white. *Philippine waters, Indonesia, Australia (Queensland). Also reported from the Persian Gulf.*

Chlamys (Chlamys) quadrilirata (Lischke, 1870)
pl. 24 fig. 7

Syn. *P. pertenuis* Dunker, 1879

Small, thin, only slightly convex, equivalve. Ears very unequal. Valves carry 10-12 radial, elevated, ribs, each made up from a number of narrow riblets. Colour usually light purple, paler near the umbo; sometimes yellowish or brown with vague concentric white bands. Inside white or cream with some lavender. *Japanese waters.*

Chlamys (Chlamys) reticulata (Reeve, 1853)

Oval, somewhat oblong shell, not very convex, about equilateral and equivalve. Both valves carry 22 smooth and regular ribs. Ears large and unequal. Colour off-white with patches of reddish-purple, showing arched lines across the ribbing. *Persian Gulf area.*

Chlamys (Chlamys) rubida (Hinds, 1845)
pl. 7 fig. 4

Small to medium size, convex, right valve less so than left one. The latter has approx. 25 usually smooth, flat ribs, with spiny riblets in interspaces. Right valve paler in colour and smoother ribs in pairs. Anterior ears much larger than posterior. Colour varies from yellow, orange, rose, red, violet to purple. Seldom white or mottled. Inside shiny off-white. *Bering Sea to California, North Japan.*

Chlamys rubida jordani (Arnold, 1903)

Syn. *P. kincaidi* Oldroyd, 1920

Smaller than *C. rubida* ss., left valve ribs smoother, more reticulations over entire shell surface. *Deep water Alaska to California.*

Note. Some authors consider *C. jordani* an intermediate between *C. hindsii* and *C. kincaidi.*

Chlamys (Chlamys) rubromaculata (Sowerby II, 1842)
pl. 7 fig. 8

Small, oval-shaped, equivalve, compressed shell with unequal ears and 18 ribs, scaly, particularly near the ventral margins. As the name indicates, with deep red spots; these are mainly on the ribs in the centre of the valve. *Northwest region of Indian Ocean.*

Chlamys (Chlamys) ruschenbergerii (Tryon, 1869)
pl. 7 fig. 5, 5a

Medium to large shell, slightly elongate, rather thin and slightly convex. Valves covered with 26-30 not very prominent rounded ribs with radiating rows of sharp, arched scales. Which are partly worn in older specimens, and a smaller smooth rib between each pair of scaly ribs. Ears about equal. Brownish-red or purple with some lighter mottling. Pale yellow or orange specimens also occur. Inside glossy, mainly cream and brown-purple with dominant white spot or area in the umbonal region of the left valve. *Red Sea, Gulf of Akaba, Persian Gulf.*

Note. See also under *C. squamosa.*

Chlamys (Chlamys) russata (Reeve, 1853)

Small, equivalve, equilateral shell with unequal ears. Valves carry about 20 narrow sharp ribs. Colour fleshy speckled with reddish brown. *Hawaiian Islands.*

Chlamys (Chlamys) sanguinea (Linnaeus, 1758)

Medium size, equivalve, rather compressed shell, with very unequal ears. Valves carry about 25 subangular scaly ribs. Colour brown, red orange or white and black variegated. *Philippines.*

Note. Somewhat like *C. varia* but ribs are more angular and thicker.

Chlamys (Chlamys) splendidula (Sowerby II, 1842)

Small to medium size shell, oval-shaped, compressed, approx. 25 rounded ribs with small erect scales. Pale, blotched, very unequal ears. Reddish in colour with pinkish-white blotches. Somewhat like *C. varia,* but according to Sowerby more compressed, regular and neat. *Eastern Indian Ocean.*

Chlamys (Chlamys) squamata (Gmelin, 1791)
pl. 7 fig. 6, 6a

Thin, small to medium size shell, equivalve. One valve flatter than the other. From 10-12 main ribs and another 6-7 intermediate ribs on both valves, with very irregular, projecting, rough scales. Cords between ribs. Unequal ears. Colour soft orange, pink, brown to purple, irregularly blotched or streaked in darker shade or creamy-white. *Red Sea, Indo-Pacific including Japan, Korea, Philippines and Tahiti.*

Note. C. tenella (Reeve, 1853) is considered a juvenile of *C. squamata.*

Chlamys (Chlamys) sugillata (Reeve, 1852)

Small shell, equivalve, equilateral, convex. Valves bear 22 sub- angular, slightly scaly ribs. Concentric striae in the intercostal areas. Unequal ears. Colour off-white with orange and olivegreen streaks. Only known from type specimen. No locality.

Chlamys (Chlamys) taiaroa Powell, 1952
pl. 7 fig. 7

Medium size, thin but strong shell with unequal ears. Comparable with another New Zealand shell, *C. dichroa*, but much more convex and ribs not as sharp, in addition *C. taiaroa* has supporting smaller riblets. Sculpture finer throughout. Colour orange or rose. Inside the same colour. *New Zealand, South Island.*

Chlamys (Chlamys) tehuelcha (D'Orbigny, 1846)
pl. 7 fig. 9

Syn. *P. darwinii* Reeve, 1853

Medium large, convex, equivalve shell with about 15-20 scaly ribs, fine radial threads on interspaces. Unequal ears. Cream background with interrupted red, orange or flesh-coloured bands and blotches on the ribs. A very variable species. *Brazil to Argentina.*

Chlamys tehuelcha forma *vanvincquii* (Bernardi, 1852) has a slightly smaller number of ribs.

Chlamys tehuelcha forma *madrynensis* (Lahille, 1906) is creamy white with 11-14 smoother ribs.
pl. 28 fig. 4.

Chlamys tehuelcha forma *spinulosa* (Bavay, 1906) is another form, with 16-18 ribs, and reported to be subfossil.

Chlamys (Chlamys) tenuigranosa (Reeve, 1853)

Small shell, only slightly convex, valves decorated with numerous thin riblets carrying very fine nodulous scales Ears very unequal. Colour: right valve cream, left one purple. *Antilles.*

Chlamys (Chlamys) teramachii Webb, 1956

Small thin, almost equivalve shell (12 mm), carrying approx. 10 broad angular radial riblets with smaller ones in the interstices on the left valve. Ornamentation of right valve radial riblets from margin to margin, central valve area being superimposed over 5 low broad obtusely scaled riblets. Entire area also marked with diagonal grooves. Ears very unequal. Colour pale red but variable, darker near the ventral edges, left valve variegated or mottled with white. Both valves creamy in umbonal areas. *Ryu Kyu Archipelago.*

Chlamys (Chlamys) testudinea (Reeve, 1853)

Medium size, moderately convex shell, with large unequal ears. Valves carry approx. 20-24 close, rounded and scaly ribs. Blotches pattern in yellow and dark brown or purple somewhat like a tortoise shell, hence its name. *Described from the Moluccas, also recorded from S. Africa, Natal.*

Chlamys (Chlamys) tincta (Reeve, 1853)
pl. 26 fig. 3, 3a

Somewhat oblong in shape, convex, small in size, translucent, very finely but irregularly ribbed and scabrous. Very unequal ears, colours varying from purple, red to pink, orange, yellow or cream. Plain coloured, mottled or streaked in white. Inside sometimes pale red. *South Africa, south and east coast.*

Note. Cox (1927) states that he cannot find a difference between this species and *C. pusio.* See also under *C. humilis.*

Chlamys (Chlamys) townsendi (Sowerby III, 1895)
pl. 9 fig. 1

Large, solid circular and convex shell, equivalve, with unequal ears. Large abyssal notch. Carries about 20 arched radial ribs on both valves. Basic colour brownish-purple with white concentric mottling, most of it in the umbonal area. Inside colour satin white with purple in umbonal region, brown-purple on hinges and margins. *Red Sea to Pakistan.*

Chlamys (Chlamys) trifida (Dunker, 1877)

Syn. *P. jickelii* Dunker, 1882

Solid, small to medium shell, inequivalve but subequilateral, slightly oblique. Valves carry 18 regular ribs, of which each has two radial grooves. Narrow interspaces and minute transverse striae. Colour brown with white variegations. *Japan.*

Chlamys (Chlamy) varia (Linnaeus, 1758)
pl. 8 fig. 1, 1a, 1b

Syn. *P. monotis* Da Costa, 1778

Small to medium, somewhat elongated, lightweight but strong, slightly convex, equivalve, unequal ears. Valves bear about 30 radial rounded ribs with small erect prickly scales. A great number of colour variations from white, orange, brown to purple with irregular concentric white or cream markings. Right valve often a shade lighter in colour than the left one. Inside glossy, same but lighter colour than the exterior. When young most Pectinidae species are attached by byssus and become free-living when adult; however *C. varia* retains this attachment life-long. *Eastern Atlantic from Norway to Cape Verde Islands and the Mediterranean.*

Note. Gmelin, 1791 described many colour forms, such as *flammea, versicolor, violacea* etc. Jeffreys (1863) named *purpurea,* Locard (1866) added *rotundata* and *strangulata.*

Chlamys (Chlamys) vestalis (Reeve, 1853)

Small, somewhat elongated, only slightly convex, near equivalve. Ears equal. Valves carry 18 closely aligned rough riblets and rough interspaces. Colour reddish with pink spots, yellowish around the umbones. *West Indies.*

Chlamys (Chlamys) weberi Bavay, 1904
pl. 24 fig. 8

Syn. *C. natalensis* E.A. Smith, 1906

Small inequivalve shell. Lower valve more convex than the upper one. Both valves ornamented with 35-40 distinct but tiny, narrow, rounded riblets with relatively wide interstices. Specifically near the margins numerous fine concentric grooves. Unequal ears. Upper valve off-white with some black blotches, each of which is surrounded by some white colour. Lower valve off-white with brown blotches. *South Africa.*

Note. Like *C. tincta.*

Chlamys (Chlamys) wilhelminae Bavay, 1904
pl. 24 fig. 9

Small, thin somewhat translucent, equivalve, oblique with very unequal ears. Valves have numerous very closely arranged striae which on the right valve become slightly nodulous near the anterior margin. Upper valve white with a few violet spots, lower valve white. *Moluccas, New Caledonia.*

Note. See also *C. marshallensis.*

Chlamys (Chlamys) zeelandona (Hertlein, 1931)
pl. 8 fig. 2,2a

Syn. *P. imparicostatus* Bavay, 1905

Small to medium size, somewhat elongated equivalve shell carrying approx. 10-12 main ribs. Finer ribbing in the interstices. Ribs have fine thorns. Very uneven ears. Colour off-white or pinkish background with purple radial stripes on the main ribs of the left valve. Usually dark bands in the intercostal spaces of the left valve. *New Zealand, South Island.*

Note. Probably a colourform of *C. zelandiae.*

Chlamys (Chlamys) zelandiae (Gray, 1843)
pl. 8 fig. 3

Syn. *P. multicostatus* Reeve, 1853

Small to medium size shell, moderately convex with 18-20 slightly scaly main ribs and additional secundary ribbing. Very unequal ears. Colour usually uniformly white, yellow, orange, red or purple. Sometimes with darker radial stripes. *New Zealand, North Island.*

Note. See also under *C. dieffenbachi.*

subgenus ANNACHLAMYS Iredale, 1939

Type species: *Pecten leopardus* Reeve, 1853.

Chlamys (Annachlamys) leopardus (Reeve, 1853)
pl. 9 fig. 6

Medium to large, fan-shaped, mat shell, with subequal ears. Each valve has from 15-20 rounded radial ribs. Interspaces wider than the ribs. Shell has numerous fine but rough concentric ridges or grooves right across the total surface. The convex valve is white, the flatter one pinkish white with ribs coloured orange, rose or tan, often interspersed with white blotches, the whole resulting in rays of colour. Inside of this species glossy white with pink or bright orange or red blotches in a regular fashion forming a semicircle on the interior ribs, strongest near the ventral margins and near the auricles. *Australia: Queensland and N. Australia.*

Note. Some authors consider *P. solaris* Born, 1778 a synonym of *C. (A.) leopardus.* One might be inclined, however, to associate this shell rather with *C. (A.) macassarensis.*

Chlamys (Annachlamys) macassarensis (Chenu, 1845)
pl. 9 fig. 7

Syn. *P. solaris* Sowerby, 1842 (non Born, 1778)

Medium size, thick, roundish, equivalve, only slightly convex shell. The valves carry about 18 smooth ribs. Small equal ears. Colour brick-red with irregular, mostly concentrically arranged, streaks, lines or bands in cream and red. Inside soft yellow, sometimes with a tinge of orange and a white central area. *Indonesia; Moluccas (named after Makassar on Sulawesi = Celebes).*

Note. Chlamys (A.) reevei (Adams & Reeve, 1850) is very close and hard to separate from *C. macassarensis* both in shape and in colours. It is considered a forma. **pl. 9 fig. 8.**

subgenus BELCHLAMYS Iredale, 1929

Type species: *Pecten aktinos* Petterd, 1886.

Chlamys (Belchlamys) aktinos (Petterd, 1886)
pl. 8 fig. 5

Syn. *P. bednalli* Tate, 1886
P. pulleineanus Tate, 1886

Small to medium, slender, slightly oblique, lightweight, depressed shell. Unequal ears, radially folded. Valves have about 10 angular scaleless main ribs and smaller ones with fine scales in the interspaces. Growth lines showing. Various colours such as cream, orange, stone-brown or purplish, sometimes with blotches or irregular lines in cream colour. Usually paler towards the umbones. Shell somewhat translucent, thus colours vaguely visible through the inside. *Shallow depths on Western and Eastern Australian coasts, including Tasmania, which is the type locality.*

Note. Often spelled *"atkinos"*. This shell is not unlike *C. asperrima* but smaller and differs by the small posterior ear, the sculpture, and is more elongated.

subgenus CORALICHLAMYS Iredale, 1939

Type species: *Coralichlamys acroporicola* Iredale, 1939.

Chlamys madreporarum (Petit in Sowerby II, 1842)
pl. 24 fig. 10

Syn. *C. acroporicola* Iredale, 1939

Lightweight, small, oblique, irregular shell. Rather flat with numerous vaguely scaled riblets or striae crossed by fine concentric lines creating a lattice pattern. Unequal ears. Left valve has cream background with radial brown, or rose, interrupted markings. Right valve cream or white. Inside white with reddish margins. Shell usually attached to coral growth. *Indo-Pacific.*

subgenus CRYPTOPECTEN Dall, Bartsch & Rehder, 1938

Type species: *Cryptopecten alli* Dall, Bartsch & Rehder, 1938.

Chlamys (Cryptopecten) alli (Dall, Bartsch & Rehder, 1938)

Small, equivalve, thin, fan-shaped shell. Both valves have approx. 20 angular ribs and small serrated cords in intercostal spaces. Ears about equal. Colour white with red speckles or streaks. *Hawaiian Islands.*

Chlamys (Cryptopecten) bernardi (Philippi, 1851)
pl. 23 fig. 3, 3a

Syn. *C. nux* (Reeve, 1853)

Rather small, equivalve convex shell, valves carrying about 15-20 radial ribs which are actually three-ridged. Intercostal areas scabrous. Unequal small ears. Colour is variable with brown, rose or purple and relatively large patches of cream or yellowish tints. *Thailand, Indonesia, Japan, Central Pacific. Deep water.*

Chlamys (Cryptopecten) bullata (Dautzenberg & Bavay, 1912)
pl. 23 fig. 6

Small, thin shell, somewhat oblique, convex but, compressed, equivalve with unequal ears. Both valves bear 20 ribs with erect scales, which join at the top and form rows of tiny nodules. Grooves concentrically lamellose. Colour of top valve off-white with rose blotches, bottom valve off-white on the outside, inside, decorated with small rose spots. *Sulu Archipelago. Mentioned from Natal by Barnard (1964).*

Chlamys (Cryptopecten) oweni (De Gregorio, 1884)
pl. 8 fig. 4

Syn. *P. pictus* Sowerby II, 1842 (non Da Costa, 1778)

Small to medium, equivalve with subequal ears. Valves carry approx. 12 round ribs with deep intercostal grooves and radial striae. Colour cream with pink or burgundy flames or streaks. *West Pacific.*

Note. Quoting Reeve (1852) : "The white parts of the ribs of the left valve are irregularly wrinkle-crenated".

Chlamys (Cryptopecten) smithi Sowerby III, 1908

Small, suborbicular, flattish, equilateral. Right valve moderately convex. Left valve only slightly convex. Both valves carry about 21 radial ribs. Fine concentric scabrous sculpture on ribs and interstices. Unequal ears. Colour yellow with reddish and white variegations. *Mauritius*

Chlamys (Cryptopecten) vesiculosa (Dunker, 1877)
pl. 8 fig. 6, 6a

Syn. *P. hastingsii* Melvill, 1888

Small to medium size, extremely convex, very solid species. Valves carry from 14-18 rather smooth flattish ribs and very distinct growth lines. Unequal ears. Colour cream to yellow, orange, reddish brown or purple, large areas of the valves blotched with darker brown or purple, sometimes lavender concentrated near the umbonal areas. Inside right valve usually white, inside left valve white lavender in hinge and umbo regions. *Japan, Indonesia.*

subgenus HINNITES Defrance, 1821

Type species: *Ostrea crispa* Brocchi, 1814 (fossil).

Chlamys (Hinnites) adamsi (Dall, 1886)

This shell is only known by a few valves and as coming from *St. Vincent, West Indies.*

Chlamys (Hinnites) corallina (Sowerby I, 1827)
pl. 9 fig. 2

Irregularly fan-shaped shell, sculptured with rough scaly ridges and frills. Colour red to orange-red. Inside light brown with purple, often with some veinous network. *East Africa, but also reported from Japan.*

Chlamys (Hinnites) ercoliniana (Cocconi, 1873)
pl. 9 fig. 3

Syn. *H. absconditus* (P. Fischer in Locard, 1898)

Medium size, thin but strong shell, inequivalve, rather flat left valve somewhat more convex, has many thin spiny ribs with minute riblets in interspaces, the (tiny) spines pointing towards the umbo. Right valve usually attached to rocks, colour yellow. Left valve also yellow and shows some irregular rays or markings of a reddish colour. Equal ears. *Deep water Cape Verde Islands and coastal areas on the opposite African mainland from Senegal to Angola.*
One of the rarest of this subgenus. *C. (H.) ercoliniana* forma *submutica* (Locard, 1898) has a truncated valve shape.

Chlamys (Hinnites) gigantea (Gray, 1825)
pl. 9 fig. 4

Syn. *P. multirugosus* Gale, 1928

Very massive, assymetric shell, superficially more like an *Ostrea*. Grows to 200 mm. Inequivalve. Only young ones are free swimming, adults are attached to rocks or other large objects. Brown upper valve has many shallow ribs covered with tiny scales or thorns. Typical for this shell is the large violet or purple spot inside. Juvenile specimens are often confused with a *Chlamys*, make sure by looking for the purple blotch near the hinge area. *Pacific, Aleutian Islands to Mexico.*

Chlamys (Hinnites) pasca (Dall, 1908)

Small, light, equivalve shell. Described by Dall from only one (left) valve, which has approx. 14 equal radial riblets and is decorated with a very fine, scaly, reticulated sculpture over the entire surface. Ears unequal. Colour rose and white with a few deep red radial bands. *Collected by the "Albatros" expedition at Easter Island.*

Note. W. H. Dall's comments on this shell: ".....it seems to belong to the group of species called *Hinnites*. I have... the opinion that the different species of *Hinnites* are more intimately related to various groups of *Pecten* than they are to each other .... The 'genus' *Hinnites* may be regarded as composed of 'sports' from the *Chlamys* group of *Pecten*".

Chlamys (Hinnites) pusio (Linnaeus, 1758)
pl. 9 fig. 5

Syn. *O. miniata* Born, 1778
O. sinuosa Gmelin, 1791
P. spinosus Brown, 1827
H. irregularis Deshayes, 1832

Small to medium shell, symmetrical when young. In adult life it becomes often very distorted or limited in its growth pattern, since attaching itself permanently by its byssus in nooks and corners between or onto rocks or similar situations. The shell is somewhat elongated with 60-70 fine narrow slightly angular riblets, crossed concentrically by minute lamellae, resulting in tiny scales at the meeting points. Colour off-white, pale yellow, reddish brown, grey or lavender, with irregular white blotches or flames. *Atlantic coasts from Norway as far as Central West Africa, also from the Mediterranean.*

Note. C. (H.) isabellae (MacGillivray, 1843), can be considered as a forma of *C. pusio*, if not a juvenile of this species. *C. (H.) distorta* (Da Costa, 1778) is closely related to *C. pusio*, and by most authors considered as a junior synonym.

subgenus LEPTOPECTEN Verrill, 1897

Type species: *Pecten monotimeris* Conrad, 1837.

Chlamys (Leptopecten) bavayi Dautzenberg, 1900

Very small, fragile, fan-shaped shell. Has approx. 20 sharp riblets. Ears unequal. Off-white with yellow to brown blotches or streaks. *West Indies, Colombia, Venezuela, to Brazil, possibly Uruguay.*

Chlamys (Leptopecten) camerella (Berry, 1968)
pl. 25 fig. 2, 2a

Small, slightly oblique shell with convex valves. Left valve more so than right one. Both valves carry 15-18 regular radial ribs. Interstices smooth. Ears unequal. Colour generally greyish white background with brownish blotches or spots. Left valve darker than the right one which has fainter markings. *Gulf of California to Panama.*

Note. See *C. (L.) latiaurata.*

Chlamys (Leptopecten) euterpes (Berry, 1957)

Small, flattish, slightly oblique shell with large subequal ears, up to 19 faint ribs. Colour apricot or rose and speckled. Left valve stronger in colour. *Shallow water off central west coast of Mexico.*

Note. Related to *C. velero* but different in that it has no scales and is somewhat stronger in colour.

Chlamys (Leptopecten) latiaurata (Conrad, 1837)
pl. 10 fig. 2, 2a

Small to medium size, thin and very light shell, slightly convex, valves with 12-16 square-shaped ribs, which are met at regular intervals by tiny concentric lamellae. Hinge long for this size shell. Ears subequal. Colour yellow-brown, uniformly streaked or mottled in white and deep brown. Inside same colour. *Shallow waters Mid-California to Baja California.*

Note. C. (L.) delosi (Arnold, 1906) is considered by some as a synonym, by others as a subspecies of *latiaurata. C. (L.) camerella* is smaller and more convex.

Chlamys (L.) latiaurata monotimeris (Conrad, 1837)
Synonym *C. (L.) lautiaurata* forma *fucicola* (Dall, 1898) and *C. (L.) latiaurata* forma *tunica* (Philippi, 1844). Same size as *C. latiaurata* s.s. but with fewer ribs which are rounder and without lamellae. Colour of the almost transparent valves yellow or orange with V-shaped streaks of white and brown.

Chlamys (Leptopecten) velero (Hertlein, 1935)

Small shell with 16 scaly angular ribs, some at regular intervals of 2. Right valve more convex, left valve almost flat. Large ears. Colour of left valve is either red-brown, red or orange, with brown and white speckles. Left valve the same but paler, sometimes even white. *Deep water Gulf of California to Panama.*

Note. See also *C. (L.) euterpes.*

C. (L.) velero forma *biolleyi* (Hertlein & Strong, 1946) is a form with about 12 ribs and triangularly shaped. Colour cream and dark brown in concentric fashion. Same range.

subgenus MIMACHLAMYS Iredale, 1929

Type species: *Pecten asperrimus* Lamarck, 1819.

Chlamys (Mimachlamys) asperrima (Lamarck, 1819)
pl. 10 fig. 3, 3a, 3b

Large, circular, slightly oblique shell. Both valves convex, covered with 25-30 spiny and scaly radial ribs and smaller ones in the interspaces. Ears unequal. Usually uniformly coloured yellow, orange, red-brown or purple. *Australian coast from New South Wales, Victoria, South Australia to Tasmania.*

Note. The forma *antiaustralis* Tate, 1886 is somewhat rounder and flatter.

Chlamys (Mimachlamys) blanda (Reeve, 1853)
pl. 10 fig. 1

Like *C. asperrima* but much smaller and more ventricose. Ears very unequal. Valves carry about 22-26 radial spiny angular ribs. Usually grey or cream background and purple-brown on the ribs. Sometimes mottled. Inside glossy, pale blue with circular patch of purple. *Australia; Queensland and South Indonesia.*

Note. *C. blanda* belongs to the *C. asperrima* complex, in which it represents the taxon from *Queensland*. Other taxa in the complex are: *C. rubida* (Menke, 1843); and *C. prunum* (Reeve, 1853) which has only 18 ribs. *C. asperrimoides* Powell, 1958 from *Norfolk Island* is smaller.

Chlamys (Mimachlamys) australis (Sowerby II, 1842)
pl. 10 fig. 4; pl. 28 fig. 6, 6a

Syn. *C. polyaktinos* Ludbrook, 1955

Very similar to *C. (M.) asperrima,* in colour and dimensions, though often larger. The valves, however, are more convex, have less ribs and as a rule have more prominent scales on the ribs. Ears almost equal. Colour uniformly yellow, orange, red, purple. *Shallow to deep water, Western Australia.*

Note. This species represents the *West-Australian* taxon in the *C. asperrima* complex.

Chlamys (Mimachlamys) consociata E. A. Smith, 1915
pl. 25 fig. 3

Small oval-shaped, only slightly convex shell. Numerous narrow ribs divided in 3 riblets of which 2 with tiny spines. Unequal ears, posterior one at an angle from the valve. Colour white to pale rose or soft red, sometimes with cream speckles or streaks. *New Zealand, deep water.*

Note. Related to *C. gemmulata.*

Chlamys (Mimachlamys) cruentata (Reeve, 1853)

Syn. *P. raffrayi* Jousseaume, 1886
P. effulgens Reeve, 1853

Medium size, equivalve, equilateral. Ribs, about 27 in number, have minute scales. Purplish background, sometimes brown, greyish or red and mottled in off-white. Inside off-white or tan with deeper red or purple-brown around the edges and near the ears which are very unequal. *Shallow to deeper water northern parts of Queensland, Indonesia, the Philippines, and Indian Ocean.*

Note. See *C. (M.) lentiginosa* (Reeve, 1853).

Chlamys (Mimachlamys) deliciosa (Iredale, 1939)

Small, convex shell with rough sculpture. The valves bear approx. 22 prickly scaled ribs. Strong hinge. Unequal ears. Colour usually uniform pale orange or brown some concentric bands in darker shades, sometimes mottled. *Australia, Great Barrier Reef.*

Chlamys (Mimachlamys) dissimilis (Montrouzier, 1858)

Syn. *P. serratus* Sowerby II, 1842 (invalid)

Small to medium size, oblique, not greatly convex. Unequal small ears. About 10 main ribs with fine scales and many finer riblets. Colour yellowish with red-brown to black irregular spots. Sometimes with irregular white lines. Ribs lighter in colour. *Philippines, New Caledonia.*

Note. Like *C. squamosa* but is not as flat, has smaller ears and scales are much finer.

Chlamys (Mimachlamys) ellochena (Iredale, 1939)

Medium size shell, left valve more convex than right one, which is rather compressed. Shell has 20 ribs covered with erect lamellae, the closer to the umbonal area the smoother; unequal ears. Colour outside pale red or yellow, inside white with rose or pale yellow edges. *In shallow water, N.E. Australia.*

Chlamys (Mimachlamys) funebris (Reeve, 1853)
pl. 11 fig. 2, 2a

Medium size, rather flat, equivalve, somewhat elongated shell, with 20-25 fine scaly ribs on both valves. Very unequal ears. Colour rose, brick-red or brown with cream or yellow markings, though other colour varieties occur. *Northern coasts of Australia.*

Chlamys (Mimachlamys) funebris curtisiana (Iredale, 1939) is somewhat more swollen, and could be considered the subspecies from *Queensland.*

Chlamys (Mimachlamys) gavena (Iredale, 1939)

Medium size, equivalve, convex shell with 26 angulated ribs on each valve. Umbonal area smooth, scaling becoming usually more distinct towards the margins. Interspaces smooth. Valves gape somewhat at the sides. Unequal ears. Colours variable, usually purplish, slightly mottled, inside light or purplish brown when fully adult. *Queensland to New Guinea, in shallow water.*

Chlamys (Mimachlamys) gemmulata (Reeve, 1853)
pl. 11 fig. 3

Small to medium sized rounded shell, thin, not very convex, with numerous scabrous riblets or threads. Unequal ears. Colour varies from uniformly off-white to yellow, orange, rose, reddish dots on a lighter background are no exception. *New Zealand.*

Note. A forma from *Stewart Island, Foveaux Straight*, for a long time called *C. (M.) gemmulata radiata* (Hutton, 1873), (non *radiatus* Bosc, 1801) has been renamed *C. (M.) suteri* (Hertlein, 1933.)

Chlamys (Mimachlamys) gloriosa (Reeve, 1853)
pl. 11 fig. 4, 4a-c

Large, strong, subcircular, convex shell. Both valves with about 22-25 rough ribs and riblets and concentric rows of minute lamellae. Unequal ears. Colours mostly uniform from lime to yellow orange, brown, magenta or deep purple. Sometimes mottled in paler tones. Colours inside as outside but softer. *Distribution over a wide area: Indian Ocean, West Pacific from Japan to the northern parts of Australia, probably also Indonesia, and other islands further out in the Pacific.*

Note. Very much like *C. (M.) crassicostata.*

C. (M.) subgloriosa (Iredale, 1939) is a local form from *Queensland, Australia*, mainly differing by often having no intercostal riblets.

Chlamys (Mimachlamys) instar (Iredale, 1925)

Small to medium size, semi-orbicular shell, convex, with approx. 20 distinct main ribs, carrying flat, not erect, scales. Further secondary ribs in the interspaces, and sculpture between the ribs. Unequal ears. Colour varies generally cream to tan, inside purple with large white muscle scar. *Southern Australia and Tasmania.*

Note. Generally similar in appearance to *C. asperrima,* which shell however has erect lamellae.

Chlamys (Mimachlamys) lentiginosa (Reeve, 1853)
pl. 11 fig. 1, 1a

Small to medium shell, somewhat oblong, not very convex, equivalve and almost equilateral. Unequal ears. Valves bear approx. 22 scaly angular ribs. Colour off white with brown or blackish speckles. Umbonal area purplish. Interior shade of white. *West Pacific.*

Note. By some authors considered a colour form of *C. (M.) cruentata.*

Chlamys (Mimachlamys) crassicostata (Sowerby II, 1842)

Syn. *P. nobilis* Reeve, 1852

Medium to large solid convex shell, with large number of strong rounded scaly ribs. In many rich colours, either unicolour in lemon yellow, warm yellow, red, brown or purple, or mottled, sometimes forming concentric bands. Unequal ears. *From Japanese water, also from Indian Ocean.*

Note. Has been considered a subspecies of *C. senatoria,* but because of its characters, it deserves specific status.

Chlamys (Mimachlamys) scabricostata (Sowerby III, 1915)
pl.11 fig. 5

Medium size shell. Convex, equivalve. About 20 ribs with large frilly scales. Unequal and scaly ears. Flesh-coloured, sometimes speckled, with some orange near umbones. Scales pale lavender. Inside glossy with pink edges. *West Australia.*

Chlamys (Mimachlamys) senatoria (Gmelin, 1791)
pl. 11 fig. 6, 6a

Syn. *Ostrea citrina* Gmelin, 1791
Pecten norvegica Röding, 1798
Pecten aurantius Lamarck, 1819
Pecten florens Lamarck, 1819
Pecten miniaceus Lamarck, 1819
Pecten similis Baird, 1873
Pecten (Chlamys) senatorius var. ex col. *articulata* Dautzenberg & Bavay, 1912

Medium size shell, slightly higher than wide, sub-oval, rather thin, about equivalve, convex, sculptured with 20-24 elevated, rounded ribs with concentrically arranged small erect scales, sometimes only visible at the margins. Spaces between the ribs as wide as the ribs themselves with obsolete growth striae. Ears unequal, byssal notch prominent. Colours vary from yellow, orange, red to purple often with white spots or blotches. Interior same colour scheme but paler. *Red Sea, Indian Ocean, Japan.*

C. (M.) senatoria porphyrea (Gmelin, 1791) seems to be a subspecies mainly *occurring in the Red Sea area, though also recorded from E. Africa, Madagascar and Mauritius.*

Note. The many synonyms represent colour formae. *C. (M.) senatoria* forma *lilacina* (Dautzenberg & Bavay, 1912) is just another colour form. See also under *C. farreri* and *C. ustulata.*

Chlamys (Mimachlamys) sentis (Reeve, 1853)
pl. 12 fig. 1, pl. 26 fig. 4, 4a-b

Small to medium, oblique shell with a great many tiny scaly ribs (up to approx. 50) of different dimensions. Moderately convex, equivalve, very unequal ears. Bright colours sometimes white, often orange, red, deep red, brown or purple either uniformly or mottled. Inside white. *Transatlantic to Brazil and the Caribbean, in shallow waters.*

Note. Somewhat similar but not *C. mildredae* or *C. ornata.*

Chlamys (Mimachlamys) squamosa (Gmelin, 1791)
pl. 12 fig. 2

Syn. *O. anonyma* Gmelin, 1791
O. sulphurea Gmelin, 1791
C. linquafelis Röding, 1798
P. larvatus Reeve, 1853
P. smegmatus Dall, 1926
C. ootanii Azuma, 1960

Medium sized, thin, not very convex, even flattish, somewhat oblique. Right valve, the more convex, carries about 20 regular ribs with numerous fine scales. Left and flatter valve carries approx. 12 low radial, scaled ribs and fine radial cords in between. Generally very various in sculpture. Scales often completely or partly worn away. Ears unequal. Colour from cream to orange, brick-red and purplish. Usually with speckles, zigzags or flecks, mainly on the ribbing, forming vaque irregular bands in lighter shades. *Japan, West and South Pacific (New Caledonia). Also recorded from West Australia, Mauritius, Madagascar, South Africa, Red Sea.*

C. sauciata (Gmelin, 1791) and *C. ruschenbergerii* (Tryon, 1869) from the *Red Sea* are the same as *C. (M.) squamosa* forma *decoriata* (Jousseaume in Lamy, 1928) from *Arabian waters.*

Chlamys (Mimachlamys) ustulata (Reeve, 1853)

Although described by Reeve as a distinct species, it appears that this shell is just another local Australian, or at least a local South Pacific forma of *senatoria.*

subgenus MIRAPECTEN Dall, Bartsch & Rehder, 1938

Type species: *Pecten mirificus* Reeve, 1853.

Chlamys (Mirapecten) mirifica (Reeve, 1853)
pl. 12 fig. 3, 3a

Small to medium size, thin, rather flat, triangular shell. Right valve more convex than the other. Both valves have about 8-10 distinct ribs or rather folds with deep and wide interstices. Ribs of some shells somewhat angular, in others more rounded. Ribs carry thorn-like curled scales at regular distances. Entire shell covered with fine radial threads, sometimes worn but usually distinct near margins. Ears, small and equal, carry these thorn-like scales on the very top. Colours vary from all cream, yellow, red to purplish. Sometimes mottled. Inside colour as outside. *Deep water from Red Sea to the West Pacific (Coral Sea, Philippines and Japan).*

Note. See also *C. (M.) rastellum* and *C. (M.) thaanumi.*

Chlamys (Mirapecten) rastellum (Lamarck, 1819)
pl. 12 fig. 4,4a-b

Very flat, small shell, almost equivalve, with about 9-10 pronounced ribs carrying rows of hollow, curved scales. Ears very large and scaly on the top. Colour pink or reddish brown, sometimes mottled. Inside colours as outside. *Philippines.*

Note. "*Pecten rastellum* Lam", figured in Reeve's Conchologia Iconica, is *C. squamata.* According to Sowerby, it is distinguished from *squamata,* in being regular and smooth, having the sides of the ribs undefined and the scales small in proportion. *C. rastellum* is very close to, and hardly distinguishable from *C. mirifica.*

Chlamys (Mirapecten) thaanumi (Dall, Bartsch & Rehder, 1938)
pl. 12 fig. 5

Small shell, left valve rather flat, other more convex. Valves have 5 distinct radial ridges, each bearing from 3-5 hollow thorn-like spines. Entire valves often covered with fine thin radial threads. Ears also spinose. Colour usually red with white blotches. Inside same as outside. *Hawaii, Philippines.*
Related to *C. (M.) mirifica,* hard to separate and some authors consider it to be synonymous.

subgenus PACIPECTEN Olsson, 1961

Type species: *Pecten tumbezensis* D'Orbigny, 1846.

Chlamys (Pacipecten) leucophaea (Reeve, 1852)

Small to medium, rounded, equivalve, equilateral, convex. Valves with 18 angular ribs and fine concentric striae. Ears almost equal. Colour reddish-brown. Inside brown and white. *Caribbean waters.*

Chlamys (Pacipecten) linki (Dall, 1926)
pl. 12 fig. 6

Small, moderately convex solid shell, with approx. 14 narrow ribs with broad interspaces. Minute close concentric lamellae. Ears subequal, large byssal notch. Both valve brown sparingly blotched with white or cream. *Gulf of Mexico, Florida.*

Chlamys (Pacipecten) tumbezensis (D'Orbigny, 1846)
pl. 12 fig. 7

Syn. *P. aspersus* Sowerby I, 1835
P. sowerbyi Reeve, 1852
P. paucicostatus Carpenter, 1864

Small to medium, somewhat convex. Right valve more so than left one. Has approx. 14 arched though somewhat angular ribs, with wide intercostal spaces. Left valve has approx. 14 ribs also, but rounder. Unequal ears. Colour right valve white with sometimes a touch of colour on the top side of the ribbing. Left valve varies from orange, tan, reddish, grey to charcoal, lighter shades between the ribs, sometimes with white mottling. Inside white. *Gulf of California to Peru.*

Note. C. (P.) indenta, Li, 1930 and *C. (P.) splendens* Li, 1930, are placed in the synonymy of *tumbezensis*, although some authors mention them as subspecies.

subgenus SCAEOCHLAMYS Iredale, 1929

Type species: *Pecten lividus* Lamarck, 1819.

Chlamys (Scaeochlamys) cookei Dall, Bartsch & Rehder, 1938
pl. 25 fig. 4, 4a

Syn. *C. albolineata* (Ostergaard, 1928) non Sowerby II, 1842

Small to medium shell with unequal ears. Right valve carries approx. 50 radial cords with smaller threads in the intercostal grooves. The main cords carry, in an irregular fashion, weak scaly ornamentation, somewhat more distinct near the ventral margin. Left valve has a similar sculpture with distinct crossmarkings between the cords. Colour creamy with brown or reddish mottling. Inside cream variegated with brownish-purple. *Hawaiian waters.*

Note. Close to *C. (S.) irregularis.*

Chlamys (Scaeochlamys) irregularis (Sowerby II, 1842)
pl. 12 fig. 8

Syn. *P. cuneatus* Reeve, 1853
P. limatulus Reeve, 1853
P. multicolor Melvill & Standen, 1907
C. midwayensis Habe & Okutani, 1968

Small to medium size, only slightly convex, thin, irregularly oblique shell, with numerous somewhat scaly thin ribs and cords. Unequal ears, distinct byssal notch. Exterior colours vary from cream, yellow, orange, brick-red, brown to purple, sometimes mottled or with a yellow area near the umbones. Inside glossy white or lavender with dark purple patch near the umbones. *Indian Ocean, Australia, Indonesia to Japan and Central Pacific.*

Chlamys (Scaeochlamys) livida (Lamarck, 1819)
pl. 12 fig. 9

Medium size, lightweight, convex shell, sometimes distorted in shape. Right valve has from 12-20 rather smooth ribs with small scales when older, usually paler in colour than the left valve. Left valve, which is more convex, has less main ribs but large fringed scales. Colour dull, very variable from uniform white, yellow, orange, red, brown, purple or mottled. Inside glossy, colour following the exterior one. Auricles large for this size shell, but unequal, some with many ribs, others with one or no riblets. *Australia, in shallow waters.*

Note. *C. (S.) foliacea* (Quoy & Gaimard, 1835) can be considered a form of the typical *livida*. The same applies to *C. (S.) tegula* (Wood, 1828), this shell may have some more and thicker scales. *C. (S.) peroniana* (Iredale, 1939) is the *Queensland* form of *livida*. When studying large series it shows that they flow into each other and borderlines of distinct variance can hardly be determined.

subgenus SWIFTOPECTEN Hertlein, 1936

Type species: *Pecten swiftii* Bernardi, 1858.

Chlamys (Swiftopecten) swiftii (Bernardi, 1858)
pl. 13 fig. 1

Syn. *P. tigerrinus* Yokoyama, 1920

Large, thick, solid, elongated, convex shell. Has approx. 5 low, broad ribs which carry rather large, widely-spaced knobs. Both valves covered with many fine radial cords. Unequal ears. In beautiful pastel colours, white, pink to carmine or soft violet, often with concentric reddish or yellow lines. Sometimes orange or speckled in umbonal area. Lower valve usually pale-coloured towards white. Interior glossy white. *Japan.*

Note. According to Zolotarev (1984) observations have been made of the growth marks of many bivalves. His studies seem to enable the determination of age and life span within certain limitations. Calculations showed that *C. swiftii* has an approx. life span of 13 years.

subgenus TALOCHLAMYS Iredale, 1929

Type species: *Chlamys famigerator* Iredale, 1925.

Chlamys (Talochlamys) famigerator Iredale, 1925

Small, only slightly convex, almost equivalve shell with unequal ears. Approx. 24 radial ribs with tiny scales; ribs, however, often remain smooth. Colours vary from yellow to orange, rose or purple with speckles or blotches in lighter shades or white. Inside similar coloration but much paler. *Southern East and West Coasts of Australia.*

subgenus VEPRICHLAMYS Iredale, 1929

Type species: *Chlamys perillustris* Iredale, 1925.

Chlamys (Veprichlamys) challengeri (E.A. Smith, 1891)

Small, very thin, oblique, partly translucent near umbones. Radial striae minute, some growth marks. Ridges show pale orange tint on one valve, other valve has reddish streak within. *Rare shell, allegedly found during dredging operations of the "Challenger" in deep water off Sydney, Australia. Locality, however, somewhat doubtful as localities were sometimes mixed up.*

Chlamys (Veprichlamys) perillustris Iredale, 1925
pl. 28 fig. 2

Small size, thin, very oblique shell. Valves with approx. 20 narrow ribs and wide interspaces, the ribs carrying thorn-like scales. Unequal scaly ears. Colours soft orange varying to orange-brown. Inside white. *Australia: New South Wales, Victoria, South Australia, in deep water.*

subgenus ZYGOCHLAMYS von Ihering, 1907

Type species: *Pecten geminatus* G.B. Sowerby I, in Darwin, 1846.

Chlamys (Zygochlamys) delicatula (Hutton, 1873)
pl. 13 fig. 2

Syn. *C. campbellica* Odhner, 1924
C. subantarctica Hedley, 1916

A delicate, yet strong medium to large, convex shell, somewhat asymmetric. Both valves with narrow, angular, rough ribs and large interspaces. Numerous fine radial striae all over, ventral edges serrated. Subequal ears. The more convex left valve coloured a soft yellow, pink or orange, with vague red-brown lines on the ribs. The other valve paler in colour or creamy. Interior glossy white. *New Zealand.*

Chlamys (Zygochlamys) lischkei (Dunker, 1850)
pl. 6 fig. 1

Syn. *P. patriae* Doello-Jurado, 1918

Large, light, slightly convex and oblique shell, medium size with numerous fine ribs of an orange-brown colour against a general background of cream. Finer riblets between the main ribs, surface feels somewhat rough. Lower valve flatter and whiter. Inside glossy white. Small subequal ears. *Uruguay, Argentina.*

Chlamys (Zygochlamys) patagonica (King & Broderip, 1832)
pl. 13 fig. 3

syn. *P. rufiradiatus* Reeve, 1853

Medium size, orbicular, equilateral, smooth, very convex. Ears slightly unequal. Has many, up to 34, almost square riblets and fine concentric sculpture. Colour cream, yellow to tan, ribs a darker flesh colour or brown. Right valve the paler, white and purplish. *Patagonia, Strait Magallanes.*

COMPTOPALLIUM Iredale, 1939

Type species *C. pauciplicatum* Iredale, 1939

Medium sized shell, inequivalved, several compound ribs, covered with small scales.

Comptopallium radula (Linnaeus, 1758)
pl. 13 fig. 4

Syn. *Pecten argenteus* Reeve, 1853

The whole shell covered with fine grooves, feels somewhat rough, 11-13 distinct, rounded ribs. Margins finely crenulated. Ears subequal. Colour cream, light brown to dark brown. Lighter in the umbonal area. Concentric mottling in lighter cream and brown. Inside satin white, brown margins along the hinge-line. *Western Pacific, down to N. Australia, Indian Ocean.*

Comptopallium radula griggi (Webb, 1957) is a subspecies from shallow water, *Cape Leveque, N.W. Australia.* Differs by having less ribs (9 or 10) which are covered with numerous small scales or fimbriations fom ventral margins to halfway up the valves. Interstices smooth (see **pl. 13 fig. 5**).

Comptopallium spiceri (Rehder, 1944)

Medium size, equilateral, left valve slightly flatter than right one. Both valves carry 8-9 rounded, broad ribs and very fine riblets, strongest on the ribs, where some at the top are elongately knobbed, the smaller marginal ribs may bear small scaly spines. Unequal ears. Colour outside white with irregular concentric streaks of yellow and or orange, which shine through on inside. Orange margins. *South of Hawaii.*

CORYMBICHLAMYS Iredale, 1939

Type species: *Chlamys corymbiata* Hedley, 1909.

Small to medium sized shell. Nearly equally inflated valves. No secondary radial riblets. Concentric sculpture.

Corymbichlamys corymbiata (Hedley, 1909)

Small solid shell, suborbicular, equilateral, right valve very convex. Left valve less convex. Somewhat gaping. Valves bear 16-18 pronounced tripartite ribs, with rows of tiny scales or pustules. Striped or flamed in browns and oranges, mostly on a cream background. Inside white. Ears very uneven. *Great Barrier Reef, Australia, but also reported from other parts of the Southern Pacific, and the Indian Ocean, W. Australia.*

Note. Mrs. Gertrude Thornley, who studied Pectinidae, informed me: "... has a surface pattern extending through two layers, not exactly alike. When raised nodules of the top layer erode away, which they do easily, even in live shells, the shell can strongly resemble *Chlamys bernardi* (Philippi, 1851)."

Corymbichlamys iredalei (Powell, 1958)
pl. 13 fig. 6

Medium size, solid, near equivalve, rounded, only slightly convex, right valve more so than left valve. Approx. 17 radial folds with deep intercostal grooves. Each of the main ribs has 7 narrow riblets, the entire surface carries close concentric lamellae developing into scales where they meet the radial ribs. Large unequal ears. Colour right valve pale orange with brownish orange blotches, left valve ditto but in stronger salmon. Inside of valves salmon-pink. *Kermadec Islands, Norfolk Island.*

DECADOPECTEN Swainson, 1840

Type species: *P. plica* Sowerby, 1839.

Solid, medium size and somewhat elongated shell, right valve the more convex of the two, generally with only a few, often striated, ribs with large interspaces. Shell somewhat shiny, rather smooth and not spiny or scaly, although valves are covered with numerous fine striae. Inside glossy. Ears subequal.

subgenus DECADOPECTEN Swainson, 1840

Decadopecten (Decadopecten) amiculum (Philippi, 1851)
pl. 13 fig. 7

Syn. *P. flabelloides* Reeve, 1852

Small, strong, orbicular, equivalve, convex shell. Left valve more convex than right one. Valves more convex in central areas, somewhat flattened at the margins, which show fine radial grooves. Both valves have 9-10 arched ribs. Subequal ears. Colour off-white or cream mottled with chestnut brown, red or deep purple. *Indo-Pacific.*

Decadopecten (Decadopecten) plica (Linnaeus, 1758)
pl. 13 fig. 8, 8a

Small to medium size, convex. Equivalve. Shell has 6-8 large folds rather than ribs, central ones the largest. Entire valves covered with fine radial striae. Ears small and subequal. Colour varies but generally white or red with violet, brown or lighter blotches or streaks. Inside glossy white with broad brown or red rims. Hinge areas nearly always brown. *Japan.*

Note. D. striatus (Schumacher, 1817), is very much the same if not synonymous with *D. plica.* It is claimed that *D. striatus* has less radial folds, but there are intermediates, so eventually, no doubt, they will all be called *D. plica.*

Decadopecten (Decadopecten) roseopunctatus (Reeve, 1853)

Small fan-shaped, equivalve. Valves carry 5 tuberculated broad ribs or folds and numerous fine radial ridges. Unequal ears. Colour generally white or off-white, left valve with bright pink dots. *South Pacific.*

Decadopecten (Decadopecten) strangei (Reeve, 1852)
pl. 13 fig. 9, 9a

Medium size shell, moderately convex, equivalve. Prominent 5-7 ribs, with further riblets thereon and between. Unequal ears. Yellow-white, brick-red or purple, blotched in red, brown and white. Inside yellow with brown accents. *Australia, Torres Strait and Queensland.*

Note. D. edentiplicata Iredale, 1939 may be a synonym.

Decadopecten (Decadopecten) subplicatus (Sowerby II, 1842)
pl. 13 fig. 10

Equilateral, somewhat triangular, flat, small to medium size shell with radiating striae and 3-5 elevated ribs per valve. Broad equal ears. Left valve cream-coloured with brick-red or deep brown blotches. Right valve same but paler. Inside mostly brown, especially around the hinge areas. *Japan to Indonesia.*

Note. Not unlike *P. velutinus.* Also very much like *P. plica.* Could even have been described from a juvenile *plica.* Sowerby (1847), stated in his Thesaurus Conchyliorum "The obliquity of the plica on the hinge, the flatness of the shell and the slight elevation of the ribs serve to distinguish this species from *Pecten plica.*"

Decadopecten (Decadopecten) velutinus (Sowerby II, 1842)
pl. 13 fig. 11

Small, oval, flattish, near equivalve, equilateral shell with 5-6 distinct but shallow round ribs and fine radial striae on both valves. Equal ears. Colour fawn, variegated with bands of white, margins brown. Sowerby described the shell as having "a somewhat velvety appearance". *Indonesia, Moluccas.*

subgenus ANGUIPECTEN Dall, Bartsch & Rehder, 1938

Type species: *Anguipecten gregoryi* Dall, Bartsch & Rehder, 1938.

Decadopecten (Anguipecten) aurantiacus (Adams & Reeve, 1850)
pl. 13 fig. 12

Small to medium, equivalve, somewhat triangular, elongated shell. Left valve flat, right valve somewhat convex, with 12-14 rounded ribs widening towards the ventral margins. Both valves finely striated, minutely scabrous. Unequal ears. Yellow, buff, orange red or violet colour with cream and brownish irregular streaks. *West Pacific. Japan.*

•*Decadopecten (Anguipecten) lamberti* (Souverbie & Montrouzier, 1874)
pl. 14 fig. 1

Small to medium, thin, flattish, equivalve, semi-orbicular with equal ears. Valves, orange in colour, carry 30-40 broad radial main ribs each having 2-3 grooves and numerous concentric lamellae. *New Caledonia.*

Note. D.(A.) gregoryi (Dall, Bartsch & Rehder, 1938) from Hawaii, is very similar if not synonymous.

Decadopecten (Anguipecten) superbus (Sowerby II, 1842)
pl. 14 fig. 2, 2a

Medium to large size, solid, oblique shell, hardly convex, small equal ears, valves carry approx. 23 strong, smooth, rounded ribs. Colour creamy-yellow background, variegated with orange or red interrupted lines or bands, sometimes a pink shade near umbones. *Japan, Okinawa, the Philippines and Papua-New Guinea.*

EQUICHLAMYS Iredale, 1929

Type species: *Pecten bifrons* Lamarck, 1819.

Equichlamys bifrons (Lamarck, 1819)
pl. 14 fig. 3, 3a

Orbicular, large shell, strong, can grow to 140 mm. Has only a few radial ribs (7-8) but many radial riblets. Ribs more distinct on right valve. Ears almost equal. Pale pastel colours from greyish-white, even some orange, to purple. Colour on top of ribs usually stronger and forms lines. Inside purple. Albinistic shells not rare. *South and South-east coasts of Australia, in shallow waters.*

EUVOLA Dall, 1897

Type species: *O.ziczac* Linnaeus, 1758.

Lower or right valve very convex and with weak rounded ribs, upper or left valve quite flat, more heavily sculptured.

Euvola raveneli (Dall, 1898)
pl. 14 fig. 4, 4a-b

Small to medium sized, light shell with approx. 24 shallow, rounded ribs. Convex valve cream ribs with flesh colour in between. Some concentric fine sculpture on entire surface. Flat valve brick-red to mauve with white blotches. Interior white. Inside white, some colour at the margins. *Ears equal. Gulf of Mexico, West Indies, and Florida to Carolina.*

Note. The shell is somewhat similar to *P. ziczac*, but has more ribs which are rounded and not flat and is much smaller.

Euvola ziczac (Linnaeus, 1758)
pl. 14 fig. 5, 5a-b

Medium size strong smooth circular shell. Flat valve has up to 34 small ribs, and patchwork of cream with a touch of rose and tan on the ribs near the ventral edge. Often orangy near the umbo and thin zigzag or radial markings. Right valve considerably convex, carries approx. 20 very low broad ribs, which in turn have normally a fine central groove. Colour a red or burgundy with cream splashes lighter in umbonal area and sometimes dominantly cream with darker browns. Inside white with purple near margins and ears, which are about equal. *SE Coast of USA (Florida) to Brazil.*

EXCELLICHLAMYS Iredale, 1939

Type species: *Pecten spectabilis* Reeve, 1853.

Shell rather small and inequivalve. Left valve flat, right valve convex. Ears rather large and unequal. Ten to twelve elevated ribs with deep interspaces. Right valve regular ribs and left valve raised and lower ones. Colour cream with brown and red spots on the ribs.

Excellichlamys concinna (Reeve, 1853)

Small to medium size shell, somewhat triangular in shape, rather flat, left valve the flatter. Valves bear 9 rounded dominant ribs. Both ribs and intercostal spaces are finely scaled. Unequal ears. Colour white, the ribs with rose blotches. *Red Sea.*

Note. Compare with *E. histrionica*. By some authors considered to be only a geographic race from the Red Sea of the latter species.

Excellichlamys histrionica (Gmelin, 1791)
pl. 14 fig. 6

Small flat shell, somewhat elongated at posterior margin, lower valve, only slight convex, carries 12 smooth ribs. Upper valve being the flatter, carries 5 broad dominant main ribs with further slightly smaller ribs in between. Serrated edges. Relatively large, unequal ears. Basic colour off-white, ribs dotted with red or black. Inside white. *Indian Ocean, and Western Pacific.*

Excellichlamys parva (Sowerby I, 1835)

Small shell, flattish, equilateral, sub-oval, sub-equivalve. Carries 8 strong ribs on each valve and has transverse striae. Ears unequal. Mainly off-white sometimes with some small light red dots. *South West Pacific.*

Note. According to Dautzenberg & Bavay (1912) *E. parva* is only a small variety of *E. histrionica*, recorded from Mauritius as well as from the West Pacific. Other authors consider it conspecific with *E. spectabilis*. The latter is certainly related.

Excellichlamys spectabilis (Reeve, 1853)
pl. 14 fig. 7

Small to medium slightly translucent, thin inequivalve shell with heavy fluted ribs about 10 to 12 in number. One valve more convex than the other. Ribs covered with concentrically aligned ridges. The flatter valve feels roughest. Interspaces of both valves finely scabrous. Unequal ears. Basic colour cream. Tops of ribs decorated with reddish and dark brown blotches arranged in concentric fashion. Convex lower valve paler and less blotches. Inside glossy white. *N. Eastern parts of Australia, Japan, Melanesia, Indian Ocean.*

FLEXOPECTEN Sacco, 1897

Type species: *Pecten flexuosus* Poli, 1795.

Shell small to medium in size, solid, almost equilateral and equivalve. The valves are slightly convex, the right valve more than the left one. Left valve five subequal radial ribs, right valve six, covered by very thin growth striae. Ears large and almost unequal. Species occur in abundant colour varations.

Flexopecten flexuosus (Poli, 1795)
pl. 25 fig. 5

Syn. *P. isabella* Lamarck, 1819
P. flagellatus Lamarck, 1819
P. polymorphus Philippi, 1836

Small to medium, strong, rather flat, equilateral, equivalve with equal, large auricles. Left valve has 5 thick ribs. Right valve 6 more or less irregular, rounded ribs. Striae in interstices. Both valves have very fine concentric lines. Margins are crenulated. Colour cream, orange, pink, brown to dark brown with random cream or brown speckles or stripes. Left valve somewhat lighter in colour. Interior glossy white. *Mediterranean Sea and adjacent Atlantic Ocean to Cape Verde Islands.*

Note. Many forms have been named, such as: *bifidus* Locard, 1888; *biradiatus* Tiberi, 1855; *concolor* Philippi, 1836; *inflatus* Locard, 1888; *lineolatus* Locard, 1888; *pyxoideus* Locard, 1888; *undulatus* Sowerby II, 1842, referring either to colour or to shape, but, as in many cases of European pectinids, too many different names for unimportant differences of environment or otherwise.

Subspecies *F. flexuosus duplicatus* (Locard, 1888).
pl. 25 fig. 6

By some considered as described from a juvenile *flexuosus*, however, considered a valid taxon.

Flexopecten glaber (Linnaeus, 1758)
pl. 15 fig. 1, 1a-b

Syn. *Ostrea maculata* Born, 1778
Pecten nebulosus Poli, 1795
Pecten flavidulus Lamarck, 1819
Pecten spectrum Reeve, 1853
Pecten liburnicus Monterosato, 1899

Thin but solid, medium size, rather flat, equivalve, with approx. 10 wide radial folds rather than ribs and fine concentric lines. Often showing growth lines. Colour most variable. Left valve from greyish to yellow or red, deep brown, or violet, often with flames, blotches, spots or zigzag in various colours. Right valve lighter in colour or just off-white. The inside whitish with or without a vaque yellow or darker tone near ears or/and margins. Large equal ears. *Mediterranean Sea, Atlantic, Southern France.*

Note. Lamarck described a large number of taxa within this species, which I consider to be subspecies, if not just local or colour forms, such as *flavidulus* mentioned above, *virgineus, virgo, violaceus.* Gmelin named the following formae: *aurantia, depressa, flavescens, lutea, modesta.* All these names are not of importance and can be ignored.

F. glaber distans (Lamarck, 1819) is by some authors considered a valid subspecies. It differs by being more convex and has about 10 ribs rather than folds as in *glaber glaber.* In my opinion just a form.

F. glaber ponticus (Dautzenberg & Dollfus, 1889) is a distinct subspecies. Right valve very flat. Both valves have 12 folds. Large ears. Colours yellow-brown, brickred, red splashed with cream. The other valve white or at least lighter in colour. *Seems only to occur in the Black Sea.*

F. glaber jonstoni (Chiaje, 1842), is another *Mediterranean* form.

F. griseus, in this work treated as a valid species, is by some malacologists considered to be a subspecies of *F. glaber.*

Flexopecten griseus (Lamarck, 1819)

Small, equivalve, convex species with approx. 10 distinct flowing ribs with more or less equivalent interspaces and fine cords over the whole surface of the valves. Concentric striae. Ears large and subequal. Colour greyish brown, variegated in white and darker tones on left valve. Right valve more off-white to pale cream, also speckled. Inside brownish, more so near margins. *Mediterranean and adjacent Atlantic.*

F. griseus subsulcatus (Locard, 1898)

In this subspecies the shell has more ribs (approx. 12), each of which has 3 small crenulated riblets. Concentric sculpture as in main species. Byssal notch larger. Colour generally as described above.

Flexopecten miles (Reeve, 1853)

Small to medium size, equilateral, equivalve, convex shell. Valves carry approx. 10

irregular ribs, entire valve surface covered with fine ridges. About equal ears. Colour reddish or darkbrown with white edges. *Indonesia.*

Flexopecten pesanatis (Reeve, 1852)
pl. 14 fig. 8

Shell roundish, equilateral and nearly equivalve, thin, irregular. The valves are decorated with radial, elevated striae and 5 rather swollen ribs. Colour of left valve off-white with stains and patches of light brown, ribs livid, right valve white. *Indonesia, Moluccas.*

Flexopecten proteus (Solander in Dillwyn, 1817)
pl. 15 fig. 3

Medium size, thin but strong, equivalve shell with large and subequal ears. Has 5-6 large folds on both valves and fine radial striae. Concentric growth lines. Large byssal notch. Shell very similar to *F. glaber* which has, however, about double the number of ribs. Colour usually yellow-brown with cream or grey blotches or patches. Sometimes orange and purple mottled, occasionally with zebra-like marking. Inside silky white. *Deep water Atlantic, Southern France, Portugal, Mediterranean and Atlantic south to Madeira and Cape Verde Islands.*

Flexopecten proteus amphicyrtus (Locard, 1888) is slightly smaller, valves more convex than the typical.

Flexopecten proteus anisopleurus (Locard, 1888) has more folds and is more swollen (= *F. praeteritus* Bucquoy, Dautzenberg & Dollfus, 1898).
Further formae of minor significance: *fusca, lutea, picta, radiata* all by Bucquoy, Dautzenberg & Dollfus, 1898, *violacea* by Locard, 1888.

Flexopecten unicolor (Lamarck, 1819)
pl. 15 fig. 2, 2a

Sometimes considered to be only a colour form of *F. glaber.* Medium large, strong, convex shell. About 10 rounded ribs or folds on both valves. Ears subequal. Right valve rather flat. Colour uniformly yellow, orange or pink, sometimes, however, with a few white splashes and concentrically arranged light brown bands. *Mediterranean, Atlantic near Gibraltar.*

GLORIPALLIUM Iredale, 1939

Type species: *Ostrea pallium* Linnaeus, 1758.

Medium size, solid, orbicular shell with valves almost equal in shape and size, only slightly convex, with about 10-15 broad lamellose ribs. Ears unequal with fine scaly radial sculpture. Shells often of an orange-red to violet colour but variable.

Gloripallium pallium (Linnaeus, 1758)
pl. 15 fig. 4, 4a

Solid shell from medium to large, equivalve, circular, both valves moderately convex, sculptured with 14-16 strong grooved ribs, carrying rows of lamellae with series of minute scales in between. Unequal ears, margins fluted. Variable in colour from orange, light red to violet, often white in the umbonal area, decorated with wavy concentric bands or blotches in lighter colours. Interior white with orangy-red at margins and in hinge area. *Japan, Pacific and parts of the Indian Ocean. Usual in areas where coral is present.*

Gloripallium sanguinolentum (Gmelin, 1791)
pl. 15 fig. 5, 5a

Medium size shell, somewhat higher than wide. Convex. Carries approx. 8 strong, smooth, rounded ribs, rows of striae in interspaces on each valve. Shell feels rough. Colour cream with orange, red of lilac blotches mainly on the ribs. Interior white with pale pink areas. Unequal ears. *Indian Ocean, Red Sea.*

Note. Looks very much like *G. pallium* but has less ribs and distinct orange or pink margins inside. Some authors claim that *sanguinolentum* is just another form of *Semipallium vexillum.* Others are inclined to group this species under *Excellichlamys.*

Gloripallium speciosum (Reeve, 1853)
pl. 15 fig. 8, 8a-c

Small to medium size. Valves carry approx. 12 broad, scaled ribs. Very unequal ears. Colours vary but mainly orangy or yellow in tone with darker spots, usually on a whitish background. Inside white with ouside colours shining through. *Described from Okinawa but also found in the Philippines and New Hebrides.*

Note. Similar to *G. pallium* but smaller and paler in colour. Some authors consider this shell to be described by Reeve from a juvenile *G. pallium. G. speciosum* must, however, be treated as a valid species in the light of presently available material.

Gloripallium spiniferum (Sowerby I, 1835)
pl. 15 fig. 6, 6a

Syn. *P. pulcherrimus* Gray, 1839

Small, rather depressed, nicely sculptured shell. Unequal valves and ears, but equilateral. Broad, spineferous, elevated ribs, about 9 in number. Ears also scaly. Colours semitransparant, white with rose. *Described from Gambier Islands, Central Pacific.*

HAUMEA Dall, Bartsch & Rehder, 1938

Type species: *Haumea juddi* Dall, Bartsch & Rehder, 1938.

Shell medium in size, thin and obliquely suborbicular. Ears unequal. Strong regular radial ribs, crossed by very fine concentric striae. Colour of right valve white, left valve mottled with red or brown.

Haumea inaequivalvis (Sowerby II, 1842)
pl. 25 fig. 1

Small, strong, suborbicular, slightly oblique shell. Unequal valves. Upper valve very convex, lower valve much flatter. About 18-20 subangular ribs, flat interstices. Small ears.

Colour top valve grey or off-white dark brown or reddish blotches, lower valve usually cream and light reddish shades: *Northern Pacific: Philippines, Indonesia and New Guinea. Indian Ocean: Andaman Islands and Mauritius.*

Haumea juddi Dall, Bartsch & Rehder, 1938
pl. 15 fig. 7

Small, strong, inequivalve, suborbicular, moderately convex species with approx. 20 smooth, radial, rounded ribs. Small equal ears. Colour left valve streaked and blotched in various shades of red and brown on a white background, right valve off-white with reddish umbo area. Inside white, umbonal area reddish. *Hawaii.*

LISSOPECTEN Verrill, 1897

Type species: *Ostrea hyalina* Poli, 1795.

Shell small in size, fragile and transparent, slightly inequivalve, rounded in shape, subequivalve and wider than high. The external sculpture consists of faint, nearly obsolete radial ridges and obscure riblets. Microsculpture obsolete concentric striae. Ears subequal and relatively large. Species occur in colour variations.

Lissopecten hyalinus (Poli, 1795)
pl. 16 fig. 2

Small shell, oval, somewhat elongated, thin, translucent and shiny, equivalve. Valves rayed with 10-15 flat, worn riblets. Entire valve show fine concentric lines. Equal ears. Horny coloured with some red and white dots. *Mediterranean.*

Note. L. succineus (Risso, 1826) can be considered as a colour form, if not as a synonym. *West Mediterranean and Adriatic Sea.*

LYROPECTEN Conrad, 1862

Type species: *Pallium estrellanum* Conrad, 1857 (fossil).

Usually strong, convex shell. Valves corrugated by 7-14 ribs, ribs and interstices both possess fine riblets. Most shells produce larger or smaller hollow knobs on the ribs. Ventral edge strongly serrated. Ears unequal and horizontally striated.

Lyropecten antillarum (Recluz, 1853)
pl. 26 fig. 1

Syn. *P.(L.) eulyratus* Bayer, 1943

Small shell, very flat valves with 10-14 round low ribs. Ears unequal. Colour varies from white, pale yellow through orange to brown, usually with blotches or sometimes streaks in off-white. Inside glossy and cream, some touch of orange near umbo. *Antillean waters to Florida, shallow depths.*

Lyropecten aratus (Gmelin, 1791)

Syn. *P. idaeus* Reeve, 1853

Shell small, somewhat irregular in shape, basically circular, medium convex, equivalve. Right valve bears about 10 strong, almost nodulous ribs with weaker riblets in between. Concentric lines showing. Left valve also 9-10 ribs separated by groups of 4-5 striae. Ears unequal. Colour purple or with markings on the ribs. Inside reddish purple and lighter edges. *Atlantic and Mediterranean.*

Lyropecten corallinoides (D'Orbigny, 1834)
pl. 16 fig. 4

Small to medium size, strong shell, both valves convex and about equal, with 9-10 distinct ribs. Numerous strong striae with even finer ones in between on ribs and interspaces. Ribs on left valve carry small knobs in 3 to 5 rows in a concentric fashion somewhat similar to *subnodosus* and *noduliferus*. Colour reddish brown, creamish, usually concentric, blotches on both valves and knobs and finer markings towards the margins. Inside white in umbonal area to reddish with yellow towards the margins. Ears unequal. *Cape Verde - and Canary Islands, Azores, St. Helena. Also reported from Spanish Marocco, the Red Sea, E. Africa.*

Lyropecten kallinubilosus (Bayer, 1943)

Shell rather large, sub-circular, inequivalve. Unequal ears. Both valves sculptured by 14 smooth, broad ribs. Colour white, mottled and clouded with a light purplish brown. Inside white. *Florida, Gulf of Mexico.*

Lyropecten subnodosus (Sowerby I, 1835)
pl. 17 fig. 1, 1a

Syn. *P. intermedius* Conrad, 1867
P. estrellanum Conrad, 1857
P. heermanni Dall, 1898

The largest and heaviest of the West American tropical species, grows up to 170 mm. Thick, solid, convex equivalve and fanshaped, 10 to 11 broad radial ribs most of which have humps or knobs; the valves covered with coarse radial grooves and threads on the ribs as well as on the interspaces. Colour from cream with brownish purple lines or patches to orange, brick-red or purple. Inside usually white whith violet margins. Ears unequal, the anterior one being the largest. *Lower California to Peru in deeper water.*

MANUPECTEN Monterosato, 1872

Type species: *Ostrea pesfelis* Linnaeus, 1758.

Shell medium in size, subtriangular, higher than wide, with eight to nine rounded pronounced radial ribs. Ribs and interspaces covered by radial riblets. External shagreen microsculptrure. Ears very unequal. Species brightly coloured, the right valve lighter than the left.

Manupecten alicei (Dautzenberg & Fischer, 1897)
pl. 16 fig. 3

Thin, slightly oblique, small, flattish shell, almost equivalve. Both valves have approx. 9 ribs, ventral edge irregular, also have fine radial striae. Ears unequal. Colour yellow to orange, glossy inside. *Atlantic Ocean, Azores to Angola, from deep water.*

Manupecten pesfelis (Linnaeus, 1758)
pl. 16 fig. 1, 1a

Syn. *P. corallinus* Poli, 1791
P. bornii Payraudeau, 1886
P. felipes Locard, 1886

Medium size, only slightly convex, oblique shell. About 8 ribs or folds on each valve. Whole shell covered with white concentric bands or blotches. *Atlantic Ocean, from France southwards to Cape Verde Islands and the Mediterranean; in deep water.*

Note. The form *elongatus* (Born, 1778) is generally larger and longer than the typical.

NODIPECTEN Dall, 1898

Type species: *Ostrea nodosa* Linnaeus, 1758.

Shell small to large in size, almost equivalve, slightly oblique, varying from convex to very convex often flatter at the ventral margins. Valves bear 10-15 strong ribs with knobs, usually larger on left valve than on right one. Distinct radial ridges. Unequal ears.

Nodipecten fucatus (Reeve, 1853)

Rounded, small shell, thin, only slightly convex, equivalve, with 12 subnodose ribs. Entire valves but especially the interstices are striated. Unequal ears. Colour white but red to violet near the edges. *West Indies.*

Nodipecten langfordi Dall, Bartsch & Rehder, 1938
pl. 16 fig. 5

Small to medium shell, orbicular, extremely bulbous. Valves have each 8 broad radiating ribs or folds bearing large knobs. Also decorated with radial cords over the entire valve. Ventral margins fluted and serrated. Ears about equal. Colour white, bright yellow or orange, sometimes mottled, with some light carmine in umbonal areas. Left valve somewhat lighter than right one. Inside colours bright red with marginal areas golden-yellow. *Hawaii.*

Note. According to Kay (1979) there is no reason to separate this species from *N. noduliferus*.

Nodipecten magnificus (Sowerby I, 1835)
pl. 29

Is a very large, orbicular, bright red shell, nearly equivalve, valves carry about 13-14 rounded nodose, grooved, ribs. Ears unequal. Colour outside mainly vermillion red. Inside white with purplish-red margins. *A rare species from the Galapagos Islands and West Colombia; also reported from the Philippines.*

Nodipecten nodosus (Linnaeus, 1758)
pl. 17 fig. 2, 2a

Syn. *Ostrea decemradiata* Gmelin, 1791
P. fragosus Conrad, 1849

Large, heavy shell, solid, convex, about equally wide as high, unequal ears. Valves have 7-10 large coarse ribs, decorated with large hollow knobs. The whole surface of the shell is covered with numerous fine riblets. Colour varies usually uniformly yellow, orange, carmine-red, brownish-red purple, but some have a combination of cream, tan, and/or violet in concentric bands. Inside colours similar but softer than the exterior colours. Size also varies, 70-140 mm. *Gulf of Mexico, Caribbean, Florida to Brazil.*

Note. Comparable, but not to be confused, with the East Pacific *Lyropecten subnodosus*. Although it was thought that the distribution of this shell was limited to the Caribbean and the West Atlantic down to Brazil, as indicated above, *N. nodosus* was in 1955 also recorded as coming from the Azores and Ascension Island. In 1981 live specimens have been trawled from the Gabon Coast, West Africa. The latter specimens differ, however, from the typical by missing the knobs or nodules on the ribs of the right valve. It seems therefore reasonable to consider them as a valid subspecies of *N. nodosus gabonensis* Nicolay, 1981. (See La Conchiglia, September 1981)

Nodipecten noduliferus (Sowerby II, 1842)
pl. 16 fig. 6, 6a

Syn. *Pecten arthriticus* Reeve, 1853

Small to medium size, equivalve shell, moderately convex. Each valve has 8 thick rounded ribs with small nodules and many radial striae. Internally plicated at the ventral edges. Ears about equal. Colour outside left valve red with angular spots. Right valve cream. Inside cream with brown areas. *Indo-Pacific.*

NOTOCHLAMYS Cotton, 1930

Type species: *Chlamys anguinea* Finlay, 1927.

Shell medium in size, both valves convex with five pronounced radial ribs. Radial sculpture of left and right valve dissimilar. Ears very unequal. Species occurs in abundant spotted and uniform colour variations.

Notochlamys anguinea (Finlay, 1927)
pl. 16 fig. 7, 7a-c

Medium size, orbicular, only slightly convex shell. Upper valve has about 7-8 folds and is less convex than the other valve. Interspaces in some specimens wide and ribless, flesh-coloured or reddish-brown or with cream concentric bands or rays. Lower valve has the same number of folds but carries finer riblets or grooves. Colour rose shaded to off-white. Interior colours reflect those of exterior. Ears unequal. *Southern and western parts of Australia.*

Notochlamys hallae Cotton, 1960

Slightly inequivalve, somewhat elongated, medium size, about 9 vague radial ribs on both valves and many riblets in between. Large byssal notch. Unequal ears. Colour white in umbonal section changing to light violet in centre and darker violet towards the ventral margins. Inside yellow-brown blotched with lilac. *South Australia; Largs Bay.*

Notochlamys tasmanica (Adams & Angus, 1863)
pl. 18 fig. 1

Syn. *Pecten mariae* Tenison-Woods, 1875

Medium size shell, convex but compressed, equivalve, unequal ears. Upper valve with 7 ribs, split up radial grooves. Also radial cords in interspaces. Variable colours all orange, flesh-coloured, red-brown or violet with cream mottling. Inside purplish. *Tasmania and South Australia.*

PALLIOLUM Monterosato, 1884

Type species: *Pecten incomparabilis* Risso, 1826.

Shell small, fragile, sometimes transparent and both valves convex. The valves are sculptured with very fine radial striae or with a reticulate pattern. Ears are very unequal. Species occur in abundant colour variations.

Palliolum incomparabile (Risso, 1826)

Syn. *Pecten testae* Philippi, 1836

Small, lightweight, orbicular, though somewhat oblique, slightly convex, translucent shell with unequal ears. Large byssal notch. Valves smooth. Left valve straw coloured, sometimes with streaks or zigzags in orange brown or red. Right valve mainly white. *Mediterranean Sea, E. Atlantic to Azores.*

Note. P. furtivus (Lovén, 1846) may perhaps be considered a subspecies of this shell. Some authors prefer the generic name *Pseudamussium.*

Palliolum notalis (Thiele, 1912)

Tiny shell (approx. 3 mm), rounded, equilateral and about equivalve. *Antartic, rare.*

Palliolum ringnesia (Dall, 1924)

Very little information available about this small shell from *Ringnes Island, Arctic Canada.*
By some authors considered to be a *Cyclopecten.*

Palliolum striatum (Müller, 1776)
pl. 25 fig. 7, 7a

Syn. *P. rimulosus* Philippi, 1844

Thin, small, circular, translucent shell. Left valve has a great number of spiny cords (60-70) which make it rough to touch. Upper valve has a network of fine radial and concentric lines. Ears very unequal. Colour orange-brown with cream or white blotches or zigzags. *Deep water E. Atlantic and Mediterranean.*

Palliolum tigerinum (Müller, 1776)
pl. 18 fig. 2, 2a-b

Syn. *P. obsoletus* Pennant, 1777

Small but strong, equivalve shell. Surface non-glossy but silky in appearance. Left valve more convex than right one, with in many cases numerous obsolete striae. Sometimes about 5 folds or ribs show distinctly. Very unequal ears. Colour varies from white to cream, yellow, orange, pink, brick-red, brown or grey, usually a combination of several colours with stripes or splashes, sometimes zigzags. Inside creamish or brown with darker areas, reflecting the outside colours. *Deep water Atlantic from Norway to Cape Verde Islands, Iceland, North Sea and Western Mediterranean.*
P. tigerinum forma *laevis* (Pennant, 1777) has no ribbing and right valve is more convex than the typical.

PECTEN Müller, 1776

Type species: *Ostrea maxima* Linnaeus, 1758.

Shell small to large, strong, orbicular, nearly equilateral. Left valve in most species nearly flat, or even concave. Right valve convex. Both valves carry about 14-20 generally strong radial ribs, angular or round in cross-section. Ribs in turn covered by riblets and or fine cords. Some shells have fine concentric lamellae, usually on the left valve, others are practically smooth, with radial ribs very weak. Large ears, almost equal in shape and size.

Pecten albicans Schröter, 1802
pl. 18 fig. 3, 3a

Medium to large shell, and broader than high, with approx. 10 well-raised broad smooth ribs on the convex valve, which is cream or pinkish in colour. The flat valve has smaller ribs and wider interstices. Brown mottled with whitish specks. Ears almost equal. *Japan and Taiwan.*

Note. This species is larger than *P. excavatus* and has no zigzag pattern. *P. albicans* forma *naganumanus* Yokoyama, 1920 is a local *Japanese form.*

Pecten chazaliei Dautzenberg, 1900
pl. 18 fig. 4, 4a

Syn. *P. tereinus* Dall, 1925

Thin, small, circular shell, lower valve convex, upper valve flat. Both valves have approx. 19-25 shallow and narrow ribs. Ears nearly equal. Colour yellowish brown, mottled with darker brown. *Gulf of Mexico, Southern Florida, Puerto Rico, Surinam, Brazil.*

Pecten diegensis Dall, 1898
pl. 18 fig. 5

Syn. *P. floridus* Hinds, 1845

Medium to large size shell, wider than its height. Right valve convex with more than 20 flat, angular, grooved, ribs. Left valve almost flat, has about 20 rounded ribs. Total surface shows numerous very fine concentric striae. Ears almost equal. Colour, shades of tan to light brown. Left valve darker, purple brown, sometimes mottled. Inside glossy white with brown edging all around. *From California to lower California.*

Note. Is related to *P. sericeus* Hinds, 1845, but certainly a valid species.

Pecten diomedeus Dall, Bartsch & Rehder, 1938

Small to medium size shell, broader than high. Right valve convex, carries low rounded broad radial ribs tending to median grooving. Ribbing and interstices show close concentric lamellae. Colour pale brown. *Hawaiian Islands.*
Known only from the type and a few valves, all dredged at about 20 m.

Pecten dorotheae Melvill & Standen, 1907
pl. 18 fig. 6

Medium to large, solid shell, rounded in shape, equilateral, inequivalve. Right valve strongly convex. The shell carries 17-18 radial ribs, having 2-3 grooves. Left valve, concave, has also a concentrically striated surface. Ears about equal. Colour of right valve pale flesh coloured, left valve reddish brown with a few darker zigzags near the umbo. Inside of shell white. *Persian Gulf.*

Pecten excavatus Anton, 1839
pl. 18 fig. 7

Large fan-shaped shell with 9-11 broad ribs. Almost equal ears. Right valve, convex, reddish with white blotches. Left, flat valve has brown zigzag patterns on lighter background. Inside white and shiny. Some reddish colour on flat valve. *Eastern Indian Ocean, West Pacific, Japan.*

Note. Very much like *P. sinensis* but larger and less convex.

Pecten fumatus Reeve, 1852
pl. 18 fig. 8

Syn. *P. fuscus* Sowerby II, 1842
P. bifidus Menke, 1843
P. filosus Reeve, 1852
N. preissiana Iredale, 1939

Medium to large, solid shell. Approx. 14-18 broad flat ribs on the convex valve and approx. 14 narrow squarish ribs on the other, slightly concave valve. Equal ears. Colour varies cream to grey, pink or light or dark brownish-purple with white and purple markings, mainly on the ribs of the concave valve. Interior white and brown with brown margins. *West Pacific, east coast of Australia.*

Note. P. fumatus is not unlike the Tasmanian *P. meridionalis* but is smaller and more swollen. *P. byronensis* Fleming, 1955 could be a variation of *P. fumatus* but might very well be an off-shoot of *jacobaeus*. In *P. byronensis* most of the main ribs are divided in 3 or 4 riblets, whereas in *P. fumatus* the ribs are undivided. See further under *P. jacobaeus*. *P. bifidus* is the West Australian form of *modestus*, typified by Reeve as "characterised by a single groove running down the middle of the ribs".

P. fumatus albus Tate, 1887, a subspecies, is less convex, colour of the convex valve white both inside and outside and the flat valve pink-white outside with darker ribs and streaked or blotched in darker shades. *Australia: Victoria and South Australia.*

Pecten jacobaeus (Linnaeus, 1758)
pl. 19 fig. 1, 1a

Type species of *Vola* (synonym of *Pecten*). Strong and solid. One of the large Pectens and being a typical representative of the genus, the left valve is flat to concave and in this case strongly fluted, the right valve convex. Both valves bear 16-18 ribs, but flat on top, which in turn are each radially grooved forming 4 ridges. Both valves have also minute concentric striae. Left valve ribs somewhat rounder and smoother. Colour left valve reddish brown, right valve cream, sometimes slightly pink towards the umbo. Inside glossy white margins reddish brown. Ears large, about equal, with vague radial grooves. *Mediterranean, Madeira, Cape Verde Islands and NW African coast.*

Note. This is the famous St. James shell used as a badge by the pilgrims to Compostella in N. Spain and also used in heraldy. Sowerby described a *Pecten* in 1842, which he called *subviridis*. According to that author the difference is supposed to be a more convex, deep, valve and more rounded and smooth ribs. These differences are hardly sufficient to justify specific status and the shell could be considered to be not more than a synonym.

In 1955 Fleming separated a *Pecten* which he named *P. jacobaeus byronensis*. However, *byronensis* is hardly distinct and therefore it is not considered valid, being practically identical to *P. jacobaeus*. Chenu, Locard and Monterosato named many varieties, such as *albida, major, rufa*, etc., which only show minor differences in colour or size.

P. jacobaeus keppelianus Sowerby III, 1905 is a subspecies from *W. Africa and also recorded from Ascension Island*. It is much smaller than *P. jacobaeus jacobaeus*, about 50-60 mm across. Convex valve white, the flat valve beautifully coloured in strong red with white dots, mainly on the ribbing. Umbonal area cream.

A complete white form *P. jacobaeus* forma *albus* was named by Monterosato, 1878.

Pecten lunaris Berry, 1963

Medium to large, solid. Left valve more convex than the right. Approx. 20 heavy ribs. Equal auricles. Colour bright, pinkish to brown, often lighter in umbonal areas. *Sonora coast, Mexico.*

Pecten maximus (Linnaeus, 1758)
pl. 19 fig. 2

Syn. *Janira intermedia* Schumacher, 1817 (= type species of *Janira*)

Strong, thick, large orbicular shell with one convex and on flat valve. In fact the largest European *Pecten*. Large and about even ears. The valves bear 14-16 broad, rounded radial ribs and numerous fine radial grooves all over the surface. Growth lines usually showing. Upper valve slightly smaller and therefore rests in lower valve. Colour of flat valve dark red or even purplish towards the top. Often ringed or irregularly spotted. The convex valve white or cream, faint yellow, pink, brown or reddish. Inside brown with white. *From shallow to deep water European Atlantic coasts, North Sea and Mediterranean.*

Note. P. similis Sowerby II, 1842, was based on only one specimen, it is considered a forma or a synonym. There is also a white form, named *albinus* Daniel, 1883. This is a rare albino form usually found together with specimens of the typical *maximus* colours.

Pecten medius Lamarck, 1819 (non *P. medius* Suter, 1913)
pl. 19 fig. 4

Medium to large, sub-orbicular shell, with equal ears. The flat valve is rather concave near the umbone with slightly produced sides carrying 14 narrow flat ribs, squarish at the sides, and concentric lines. Colour pale yellow, pinkish or brownish with blotches or streaks.

Convex valve has 16 broad, divided, somewhat rounded but more flattish ribs. Colour off-white or brownish with spots or purple bands. *Australia, New Zealand.*

Note. This shell is related to, but is not, *P. fumatus, P. meridionalis* or *P. novaezelandicus.*

Pecten meridionalis Tate, 1887
pl. 19 fig. 3

Large, solid, smooth shell. Convex valve with 16-18 well-raised, fluted ribs. Entire valve areas covered with concentric cords. White with pinkish hues, sometimes greyish or light purple. Flat valve rose-brown with same number of ribs, but these are flatter and narrower. Inside glossy purple-brown with large white spots in adductor muscle area. Equal ears. *Main locality Tasmania.*

Note. P. fumatus is a related species from Queensland and New South Wales and somewhat smaller but more convex.

Pecten novaezelandicus Reeve, 1852
pl. 19 fig. 5

Syn. *P. laticostatus* Gray, 1835
P. medius Suter, 1913 (non Lamarck, 1819)

Heavy large, circular shell. Right valve very convex with 14-18 distinct and prominent rounded ribs with narrow interspaces. Colour rose, brown or purple, sometimes with slight mottling. Left valve flat, ribs narrower and square, larger interspaces, white, flesh coloured or reddish brown, sometimes mottled. Interior white, sometimes with chocolate brown areas. Ears almost equal. *New Zealand and Australia. P. novaezelandicus* forma *rakiura* Fleming, 1951, represents a form, it is somewhat paler. *It is from Stewart Island, New Zealand.*

Note. By some authors considered to be a synonym of *P. fumatus.*

Pecten sericeus Hinds, 1845
pl. 19 fig. 6

Medium to large size, round shell, about 20-24 sharp ribs on both valves. Left, convex valve cream rose, or tan-coloured. Ribs stronger in colour than the background. Right valve flat or even slightly concave and paler in colour. Ears about equal. *Deep water, California to Mexico and Peru.*

Note. Somewhat similar to *P. diegensis* of which the flat valve is slightly convex.

Pecten sinensis Sowerby II, 1842
pl. 19 fig. 7

Syn. *P. puncticulatus* Dunker, 1877

Medium size, equilateral, circular, solid. Right valve convex, has about 12 flat broad ribs, numerous radial striae. Flesh-coloured, brick-red, orange or pale lavender with reddish-orange or white mottling. Ribs on left, concave, valve more squarish with large interstices. Concentrically striated with growth lines. Sub-equal ears. Inside glossy, brown to flesh-coloured with white central and umbonal area. *Japanese waters, China.*

Note. Sometimes considered a synonym of *P. excavatus.*

Pecten sulcicostatus Sowerby II, 1842
pl. 19 fig. 8

Syn. *Pecten capensis* Sowerby III, 1892

Medium to large size, orbicular, strong shell. Upper, flat, to concave valve carries 12-18 broad, radial, grooved ribs, lower valve convex, same number of ribs but broader and fine radial striae over entire surface. Concentric growth lines. Ears striated and about equal. Colour fawn or rose to purplish, mottled on the upper valve. Flat valve generally darker in colour. Inside white with broad fawn or flesh-coloured margins. *East Coast of South Africa.*

Pecten sulcicostus forma *casa* Van Bruggen, 1961

This form has no grooves on top of the ribs and in the interstices areas with exception of a few central grooves on the ribs of the convex valve. Shell slightly flatter, ribs on the convex valve also flatter than the typical. Colour lower valve dark reddish brown with paler spots. Upper valve light brown-orange with reddish brown blotches. *SE. coast Cape Province, South Africa.*

Pecten waikikius Dall, Bartsch & Rehder, 1938
pl. 20 fig. 1

Shell small to medium size. Right valve very inflated decorated with 11-15 rounded ribs, radial threads and fine concentric grooves. Left valve, concave, bears 13-14 smooth cords. Colour white with some flesh-coloured variegations. *Hawaii.*

subgenus MINNIVOLA Iredale, 1939

Type species: *Minnivola isomeres* Iredale, 1939.

Pecten (Minnivola) pyxidatus (Born, 1778)
pl. 20 fig. 2

Syn. *P. crebricostatus* Philippi, 1844

Small to medium, smooth, orbicular shell. Ears equal. Flat valve has approx. 22 narrow ribs and wide interstices, generally flesh-coloured, somewhat darker in the grooves often with brown, green or lavender areas or blotches, sometimes slightly mottled. Convex valve rather ventricose, white, cream, orange or pinkish in colour with often light mottling in brown or lavender, and also approx. 22 but wide ribs, and narrower intercostal areas. Inside generally white. *Indian Ocean, China Sea, Philippine waters south to Indonesia and Queensland.*

Note. Notwithstanding the description given for *M. isomeres* by Iredale, 1939 this so-called new Queensland species, which he used as the type species for the genus *Minnivola*, is in fact impossible to differentiate from *P. pyxidatus*. Should be considered a synonym.

subgenus OPPENHEIMOPECTEN von Teppner, 1922

Type species: *Pecten subbenedictus* Fontannes, 1878 (fossil).

Pecten (Oppenheimopecten) erythraeensis Sowerby II, 1842
pl. 20 fig. 3

Shell not unlike *P. fumatus* Reeve, 1852, but is somewhat smaller and narrower with about 15 ribs, which are generally flatter. Unequal ears. Upper valve pinkish or bluff-coloured has some red blotches, zigzags and lines. *Red Sea, E. Coast Africa.*

Pecten (Oppenheimopecten) galapagensis Grau, 1959

Although related to *P. vogdesi,* this is a different species. Medium size, large subequal ears. Right valve deeply convex and umbo strongly "incurved" and inflated. Has 15 rounded and smooth ribs with riblets in interspaces and fine growth striae. Colour mostly white with pale brown mottling in umbonal area. Left valve concave, particularly in the central area, has 14 shallow ribs, colour brown or rose mottled with lighter or darker colours, interior white with pink or brown at ventral margins. *Galapagos Islands.*

Pecten (Oppenheimopecten) hancocki Grau, 1959

Shell inequivalve, medium size, oblique, convex. Left valve rest in right one. Both valves have 16-17 strong rounded ribs and low riblets adjoining. Ears near equal. Colour right valve white and pale orange near umbo, left valve red-orange or burgund shading to brown at margins. Top of ribs grey-white. *From Cocos Island, Costa Rica, rare.*

Note. See: "Pectinidae of the Eastern Pacific, plate 57", by Grau. Nearest relative is *P. vogdesi* but quite distinct.

Pecten (Oppenheimopecten) perulus Olsson, 1961
pl. 20 fig. 4

Small to medium size shell, solid. Right valve strongly convex with approx. 22 rounded ribs. Colour purplish brown, red or white, in umbonal area. The left valve carries about 18 flatter ribs each split by 2 radial grooves. Colour brown to purple, sometimes rib tops white. Ears equal. *Panama to Peru.*

Pecten (Oppenheimopecten) vogdesi Arnold, 1906
pl. 20 fig. 5

Syn. *P. dentatus* Sowerby I, 1835
P. cataractes Dall, 1914

Medium to large, strong shell. Flat or even concave left valve smaller than the right one and fits within the other at the ventral edge. Right valve very convex and carries 20-22 rounded ribs deeply serrated like sharp teeth at the ventral margin. Left valve has about 20-22 low square ribs. Colours right valve from pale tan to brownish red, left valve in brown tones, sometimes with some concentrical lines of deep brown or purple near the umbo. Inside convex valve white with a touch of lavender on wings, which are about equal size. Left valve inside lavender with large white patch in adductor muscle area. *S. California to Panama, W. Colombia.*

subgenus SERRATOVOLA Iredale, 1939

Type species: *P. tricarinatus* Anton, 1839.

Pecten (Serratovola) asper Sowerby II, 1842

Small to medium shell. Upper valve almost flat, slightly elevated near the top, 18 ribs flattening out towards the margin. Lower valve convex, with 17-18 ribs, rounder towards margin, which is serrated. Some fine concentric striae. Interspaces narrow and level. Ears about equal. Colour upper valve mainly brick-red to brown, lower valve mostly off-white. Inside white with ruby-red blotch in umbonal area of the flat valve. Generally interstices have deeper colour than the ribs. *West Pacific.*

Note. By some authors considered synonym with *P. tricarinatus* Anton, 1839.

Pecten (Serratovola) gardineri (E.A. Smith, 1903)
pl. 20 fig. 6

Small to medium size, about equivalve and equilateral. About 18 radial, rounded ribs on each valve, which are usually striated. Colour basically cream with brown mottling sometimes dominating. Unequal ears. *Indian Ocean, Western Pacific.*

Pecten (Serratovola) tricarinatus Anton, 1839
pl. 20 fig. 7, 7a

Syn. *P. passerinus* Hinds, 1845

Small to medium shell, rather flattened, thin unequal valves, ears equal, large byssal notch. Upper valve almost flat with approx. 18 ribs, which are angular nearing the umbonal area. Top flattened forming finely serrated double angles at margins. Lower valve more convex, ribs smooth and round near margins. Fine growth lines showing. Colour top valve brown-reddish mottled with cream. Lower valve markings and bands less frequent, resulting in overall whiter appearance. Inside white. *Japanese waters, Philippines, Indonesia to New Guinea.*

PATINOPECTEN Dall, 1898

Type species: *Pecten caurinus* Gould, 1850.

Shell large and relatively thin, circular and flattish, equivalve, with up to 26 radial ribs. Right valve more convex than the left one, has broad and angular ribs. Ribs of the left valve are narrower. Subequal ears.

Patinopecten caurinus (Gould, 1850)
pl. 21 fig. 3, 3a

Very large shell, strong, nearly circular and compressed. Valves carry 17-22 low ribs. Fine concentric growth lines prominent. Left valve nearly flat, smaller than right valve and rests in the right one. Colour dirty cream to grey-reddish or purplish in colour. Right valve more convex, white ribs becoming ochre-brown in colour near ventral edge. Inside white with satin sheen. Ears subequal, large byssal groove. *Alaska to California, on muddy or sandy bottoms, deep water.*

Patinopecten caurinus yessoensis (Jay, 1839)
pl. 21 fig. 1

Generally very much like the typical specific form, but usually larger, right valve more swollen, ears larger, minor surface sculpture, more reticulated when valves are not worn. Byssal groove not so distinct. It is a subspecies *from Japanese waters.*

Note. Masuda (1963) described a subgenus called *Mizuhopecten* and used this shell as the type species. *P. poecilus* Adams, is described from a juvenile *P. caurinus yessoensis.*

PLACOPECTEN Verrill, 1897

Type species: *Pecten clintonius* Say, 1824 (Miocene).

Shell large in size, solid, without ribs and radial striae. Both valves convex, right valve smoother than left one. Ears nearly equal. Left valve coloured, the right one is white.

Placopecten magellanicus (Gmelin, 1791)
pl. 21 fig. 2

Syn. *P. grandis* Solander , 1786
P. testudinarium Röding, 1798
P. tenuicostatus Mighels & Adams, 1842 (= juvenile *magellanicus)*
P. fuscus Gould, 1848
P. brunneus Stimpson, 1851

Largest of the American pectens, grows up to 200 mm, circular in shape. Valves are only slighly convex, about equal in shape, gaping at the sides from halfway up, the opening increases towards the ears, which are subequal. Shell surface has numerous radial threads and minute scaly texture, which makes the shell feel rough, though it has the appearance of a smooth surface. The left valve is fleshy pink to brick-red on a cream base. Sometimes with darker radial bands and concentric growth lines. Right valve off-white. Inside cream and very shiny. *Deep water, Newfoundland to North Carolina, USA.* Not from the Southern parts of S. America, it's name is misleading!

Note. P. clintonius was originally described from a fossil belonging to the *Placopecten* group and named by Say, 1824. This shell is, as it were, redescribed by Verrill (1897) for a living shell of this type, comparable if not synonymous with *P. magellanicus.*

PSEUDAMUSSIUM Mörch, 1853

Type species: *Pecten septemradiatus* Müller, 1776.

Shell medium in size, almost circular in shape, fairly oblique, moderately convex, almost equivalve and equilateral. Five to ten radial ribs. Sculpture microscopic radial and concentric striae on surface. Ears small and almost equal. Colour of left valve red and brown with blotches, stripes or dots in lighter colours, often white. Right valve mostly white.

Pseudamussium davidsoni (Dall, 1897)

Small shell (about 14 mm), rounded, flat. Right valve carries numerous fine, mostly obsolete scaly riblets. Left valve carries 21 rounded riblets with rows of minute scaly subglobules, wide interstices. Small ears. Colour white. *Bering Sea.*

Note. Some authors, such as Grant & Gale (1931) and Grau (1959), suggested to assign this species to *Propeamussium* (Propeamussiidae). The situation is not yet completely clear.

Pseudamussium hybridum (Gmelin, 1791)
pl. 20 fig. 8

Syn. *P. exoticus* Chemnitz, 1795 (name rejected)
P. dispar Lamarck, 1819
P. orbicularis Sowerby II, 1842 (juv. of *hybrida)*
P. pseudamussium Sowerby II, 1842
P. loveni Dunker, 1853 (juv.)

Small to medium size shell, orbicular, not completely equivalve, thin and somewhat oblique. Only slightly convex. Left valve has very fine radial striae, almost evenly greyish-brown background with reddish-brown around the umbonal area. Right valve with concentric striae and paler in colour going towards white, inside colours: left valve light purple, right one off-white. Almost equal ears. *West Africa down to Angola.*

Pseudamussium septemradiatum (Müller, 1776)
pl. 20 fig. 9

Syn. *O. peslutrae* Linnaeus, 1758
O. triradiata Müller, 1776
P. aspersus Lamarck, 1819
P. quadriradiatus Lamarck, 1819
P. jamesoni Forbes, 1839
P. stellatus Reeve, 1853

Small to medium, fragile, equivalve, circular shell. Rather flat, with from sometimes 3 only up to 10 longitudinal folds which become wider towards the ventral edges. Edges wavy. Ears small, almost equal. Right valve white, sometimes with a reddish touch in umbonal area and on ribs. Left valve brown, purple, flesh-coloured or red with small white speckles. Inside glossy white. *Arctic, E. Atlantic to NW. Africa, North Sea and Mediterranean.*

Note. It may be mentioned that Locard named a number of formae, such as *alba, major, marmoratum.* Some authors consider *P. septemradiatum* as belonging to the subgenus *Peplum.*

Subgenus PEPLUM Bucquoy, Dautzenberg & Dollfus, 1889

Type species: *Ostrea clavata* Poli, 1795.

Pseudamussium (Peplum) clavatum (Poli, 1795)
pl. 20 fig. 10

Syn. *O. inflexa* Poli, 1795

P. fimbriatus Philippi, 1844
P. plica Stossich, 1865 (non *plica* Linnaeus, 1758)

Small to medium size shell, inequivalve, small, almost equal ears. Variable in size and shape. Both valves have about 6 ribs or folds and many radial cords. Valve margins inflexed with the cords very pronounced. Colour reddish brown with fine white speckles on left valve, cream with some red or brown patches near umbo on right valve. Inside pinkish to light violet. *Atlantic from Scotland to Cape Verde Islands and Mediterranean, in shallow to very deep water.*

Note. A number of formae has been recognized such as *depressum, griseus, inflatum* by Locard (1885); *albida* and *marmorata* by Monterosato (1875) and *elevatum* Poli, 1795. All of these are hardly worth individual description.

P. clavatum dumasii (Payraudeau, 1826)
This subspecies has the same ribstructure but a different overall sculpture. Synonym hereof is *P. costulatum* Locard, 1885.

Pseudamussium (Peplum) fasciculatum (Hinds, 1845)
pl. 20 fig. 11

Syn. *P. panamensis* Dall, 1898
P. miser Dall, 1908

Small shell, thin, broad, slightly inequivalve and oblique, slightly convex. Left valve, however, somewhat flatter. About 5-8 ribs or strong corrugations, central ribs strongest. Fine radial ridges. Colour reddish or violet, cream markings. Inside valves peach to violet. Ears unequal. *From Gulf of California to Panama in deep water.*

SEMIPALLIUM Jousseaume in Lamy, 1928

Type species: *Pecten tigris* Lamarck, 1819.

Medium size, elongated shell, has one slightly convex valve and one rather flat valve and about 9 radial folds rather than ribs, lightly frilled riblets on the folds and in the interspaces. Ears unequal.

Semipallium amicum (E.A. Smith, 1885)
pl. 26 fig. 5

Small, oblong, compressed, oblique, 8 broad rounded ribs, decorated with thin prickly scaled riblets. Deep narrow interspaces with a fine regular reticulation. Unequal ears. Colour orange-reddish, with off-white. *Tonga Islands, Tongatapu Id., off Nukalofa.*

Semipallium crouchi (E.A. Smith, 1892)
pl. 20 fig. 12

Small to medium size, oblique, elongated shell, not unlike *S. pesfelis.* Valves have 8 ribs or corrugations in radial direction. Numerous fine concentric lamellae. Entire valves radially striated. Unequal ears. Colour cream to yellow speckled with magenta. Inside cream. *Indian Ocean, Mauritius.*

Semipallium dianae (Crandall, 1979)
pl. 5 fig. 3, 3a-b

Small to medium, lightweight, oblique, only slightly convex, equivalve shell. Carries 9 elevated rounded ribs, rather wide intercostal spaces. Very unequal ears. Colour left valve pale or sometimes bright orange, or brown. Umbonal area yellow. Irregularly mottled in dark purple and white, mostly on the ribs. Right valve similar but paler. Shell transparent, ouside colours shining through. *Okinawa.*

Semipallium fulvicostatum (Adams & Reeve, 1850)
pl. 20 fig. 13

Strong, small to medium size, oblique, elongated shell, slightly convex, with 7-10 distinct ribs and sculptured with fine, sometimes prickly, radial grooves over the entire valves. Central edges finely serrated. Colour from white to pale yellow, reddish, to light and dark brown or mauve, usually with irregular speckles, blotches or streaks in white from umbo to half-way the valves. Some show darker concentric lines. Ears very unequal. *Japan, W. Pacific, Indian Oceran, Persian Gulf.*

Note. S. luculentum (Reeve, 1853) *Indian Ocean*, and *S. dringi* (Reeve, 1853) *Australia* are, according to Dautzenberg & Bavay (1912) not more than varieties of this species. *S. luculentum* has a few more ribs, *S. dringi* is usually deep purple with brown markings. Another closely related shell is *S. wardiana* Iredale, 1939, which is the type species for *Complicachlamys.* It seems to be not more than a *Queensland* form of *S. fulvicostatum.*

Semipallium imbricatum (Gmelin, 1791)
pl. 20 fig. 14, 14a

Small to medium, light, triangular shell. Lower valve convex, upper one rather flat. About 8-10 ribs which carry hollow scales or imbricated tubercles, thin striae in interspaces. Very unequal ears. Colour off-white or rose with reddish rectangular blotches. Inside yellow with some violet stains. *West Indies to Florida.*

Semipallium rapanense (Bavay, 1905)
pl. 26 fig. 2

Small shell, inequivalve, almost equilateral, only slightly convex, slightly gaping. Valves carry 9 shallow ribs divided by 2-3 grooves, with weak nodules. On the lower valve, which is the flatter, the ribs are somewhat larger. Very fine scales in the interstices. Colour yellow-reddish with small darker spots on the nodules. Described *from Rapa Island, SE. Pacific.*

Semipallium tigris (Lamarck, 1819)
pl. 22 fig. 1, 1a

Typical for this solid, flat, medium size shell is the asymmetry of the valves. On both valves about 10-11 main ribs with relatively deep interstices. Numerous fine radial ridges cover both valves. Unequal ears. Colour off-white background with purple or brown blotches forming concentric wavy bands. Inside glossy cream with much yellow on the internal ribbing. Brown spot at hinge area. Pink spot on inside ear. *Indo-Pacific down to Queensland and northern NS. Wales, Australia.*

Semipallium vexillum (Reeve, 1853)
pl. 22 fig. 2, 2a

Syn. *P. digitatus* Hinds, 1845
P . janus Montrouzier, 1858
P. zeteki Hertlein , 1935

Small to medium size, strong, circular shell, moderately convex with approx. 10 arched ribs. About equal ears. Colours vary, typical are the wavy markings in white on a background of darkbrown on one valve. Occasionally orange with white blotches. The other valve is much paler in colour. Inside white but with distinct, usually chocolate-brown areas, specially near the umbones. *Indian Ocean, West Pacific.*

Note. Bractechlamys evecta Iredale, 1939, is by some authors considered to be a synonym of *S. vexillum.* See also under *Gloripallium sanguinolentum.*

JUXTAMUSIUM Iredale, 1939

Type species: *Juxtamusium oblectatum* Iredale, 1939.

Thin, compressed, equivalve. Valves slightly gaping. Ears unequal. Sculpture of fine radial striae, nearly obsolete on left valve. Only ribbed on internal margin.

Juxtamusium coudeini (Bavay, 1903)
pl. 26 fig. 6

Small to medium, equivalve, orbicular shell. Right valve more convex than left. About 20 small ribs with concentrically arranged tiny, hardly visible lamellae. Unequal ears. Colour variable, from off-white with some brown colouring on the convex valve to chocolate brown with cream speckles. Ears lighter in colour. Inside white with brown margins near the auricles and in the muscular area. *Moluccas, New Caledonia, Philippines.*

Juxtamusium maldivense (E.A. Smith, 1903)
pl. 26 fig. 7

Syn. *J. oblectatum* Iredale, 1939

Thin shell, small, equivalve, rather flat. No actual ribs, but with or without vague folds. Both valves have fine or coarse radial striae, usually in groups of 4 or 8, right valve more so than left one, where striae sometimes become non-existing. Relatively large ears. Colour creamy with large patches or streaks of rusty or reddish brown. Right valve somewhat paler. Inside pinkish, yellow near umbones. *Indian Ocean (Maldive Islands), Red Sea, NE. Australia, New Hebrides, Fiji, Marshall Islands, Philippines.*

Note. J. oblectatum differs slightly in that its radial cords are rounder and is normally somewhat larger.

MESOPEPLUM Iredale, 1929

Type species: *Mesopeplum caroli* Iredale, 1929.

Medium size, solid, broad shell. Convex valves with 5-6 ribs and radial striae on ribs and in interstices, no lamellae or knobs on the ribs. Ventral edges fluted and with fine grooves internally. Ears sub-equal.

Mesopeplum caroli Iredale, 1929
pl. 22 fig. 4

Medium size shell, very bulbous and fluted valves. Equilateral with about 5-6 broad ribs on each valve. Both ribs and interstices covered with axial riblets or striae, and concentric grooves on the right valve. Ventral margin serrated. Ears almost equal, small and slightly spiny in the hinge area. Left valve cream with rays of crimson on ribs and cords. Right valve off-white. Inside white and glossy, slightly crimson near umbo. *Central East coast of Australia.*

Note. Mesopeplum fenestratum (Hedley, 1902) is a synonym based on a juvenile *M. caroli.*

Mesopeplum convexum (Quoy & Gaimard, 1835)
pl. 22 fig. 3

Syn. *M. kapitiensis* Mestayer, 1929

Small to medium size, solid, equivalve, but with uneven ears. Both valves convex, each with 5 broad ribs or folds gradually widening towards the ventral margins and many radial riblets or striae over the entire valve. Colour off-white with rose, orange, yellowish hue or purple. Often light brown markings between ribs. Inside glossy white with lavender. *New Zealand in deep water.*

Mesopeplum triggi (Cotton & Godfrey, 1938)
pl. 28 fig. 5

Medium size, strong, with unequal valves. Left valve shallow with 6-7 main ribs, made up by groups of smaller ribs. Interspaces wide, with riblets. Colour pinkish. Right valve more convex, also 6-7 main ribs and many riblets in between. Colour off-white. Concentric growth lines. *South and West Australia in deep water.*

VOLACHLAMYS Iredale, 1939

Type species: *Pecten cumingii* Reeve, 1853.

Shell medium in size, both valves convex with eighteen to twenty strong rounded radial ribs without scales. The interspaces are striated. The byssal notch is fairly large, the species occurs in abundant colour variation.

Volachlamys cumingii (Reeve, 1853)
pl. 22 fig. 7, 7a

Medium size, strong, convex, ears large and subequal, sometimes with white streaks. Surface bears approx. 18 smooth rounded ribs without scales or lamellae. Valves internally grooved and with fine concentric striae. Rather large byssal gape. Colour very variable, but basically cream background with orange, brick-red, brown or violet or combination thereof, blotched pattern. Some shells however, are almost white with vague yellow or rose patterns. Inside glossy white changing into soft yellow towards the edges. Pinkish spot either subtile or strong in umbonal area. *Queensland and Northern Territory, Australia.*

Note This shell is very close to *V. singaporina* (Sowerby II, 1842) which species is considered by some authors as a synonym. The ears of *V. singaporina* are more unequal. To make matters more complicated, there is a N. Australian form of *V. cumingii* which approaches very closely the Singapore form. *P. layardi* Reeve, 1853 is certainly closely related, if not identical.

Volachlamys fultoni (Sowerby III, 1904)
pl. 5 fig. 6

Medium size, circular, equivalve, thin, compressed shell. Valves carry about 17 rounded ribs, wide interspaces, no radial striae except for some indistinct ones in interspaces. Entire valves crossed by thin lamellae or growthlines. Right valve somewhat smoother than left one. Unequal ears. Colour dark purplish brown with irregular whitish markings. *South Africa, Natal.*

Volachlamys hirasei (Bavay, 1904)
pl. 22 fig. 5

Solid, medium size, rather flat, equivalve shell. Upper valve the more convex, has 13-14 flattish rounded ribs, narrow interstices. On the lower valve the ribs are smaller and interspaces wider. Ribs not very scaly. Almost equally large ears. Purple-brown to yellow-orange in colour, lighter on the ribbing, sometimes with concentrically arranged cream blotches. Lower valve paler. *Japanese waters.*

V. hirasei awajiensis (Pilsbry, 1905) (synonym *atsumiensis* Yokohama, 1920) has more pronounced ribs, 20 in number, but is generally of the same size. Purple-brown mottling with some lighter specks on creamy background. Shiny.
Another subspecies, *V. hirasei ecostata* (Bavay, 1904), is almost smooth.

Volachlamys singaporina (Sowerby II, 1842)
pl. 22 fig. 6

Syn. *P. pica* Reeve, 1853

Medium size, fan-shaped, somewhat depressed, equivalve, equilateral shell. Approx. 24 slightly elevated ribs. Unequal ears. Colour creamish background with reddish brown blotches and stripes, greyish blotches near umbones. Inside white with sometimes pinkish area near hinges. *Eastern Indian Ocean.*

Note. Somewhat like *V. tranquebaria* but flatter and with greater number of ribs. Ears, although unequal, are more proportional. Even closer to *V. cumingii.*

Volachlamys sinomarina (Hertlein, 1936)

Syn. *C. ambiguus* Bavay, 1904

Medium size shell, rather swollen, inequivalve, with crenulated edges and large subequal ears. Valves carry approx. 13 distinct angular ribs and numerous fine, low, concentric lamellae. Colour: top valve outside dark grey with white and yellowish spots, inside violet with white spots; lower valve outside off-white with touches of violet or yellow. Inside white. *China Seas.*

Note. According to L.G. Hertlein (1936) the name *Chlamys ambiguus* was already used for a fossil. Therefore he introduced the name *Chlamys sinomarina* to replace *C. ambiguus* Bavay (non Münster, in Goldfuss, 1833).

Volachlamys tranquebaria (Gmelin, 1791)
pl. 22 fig. 8, 8a

Medium size, solid, fan-shaped, swollen shell, sub-equivalve, very unequal ears, somewhat glossy. Both valves carry 18-20 ribs, smooth on top, striated at the sides. Deep interstices. Cream to rose base with relatively large brown or red-violet patches or usually horizontal, irregular bands. Sometimes orange. Inside glossy white. *Indian Ocean, Philippines, S. China, Indonesia, New Guinea, N. Australia.*

Family PROPEAMUSSIIDAE

PROPEAMUSSIUM de Gregorio, 1884

Type species: *Pecten ceciliae* de Gregorio, 1884 (fossil).

Smal, thin translucent shell. Inside and outside glossy. Valves gaping laterally near the top, both valves slightly convex and somewhat oblong. Usually smooth externally, but some species have fine striae in both concentric and radial direction. Interior has from 9-12 fine radial whitish threads from the hinge area to about the centre of the valves, occasionally yellow or orange coloured. Ears small and almost equal. No byssal sinus.

Propeamussium alcocki (E.A. Smith, 1894)

Small to medium shell, smooth valves with very fine circular striae. Approx. 5 faint radial lines on one valve. Colour creamy to light violet and iridescent. *India, Lacadive Islands, in deep water.*

Propeamussium caducum (E.A. Smith, 1885)
pl. 27 fig. 5, 5a

Syn. *P. nakazawai* Kuroda, 1932 (nom.nud.)

Small, thin and flat, gaping, somewhat elongated. Valves generally smooth on the outside, sometimes with concentric or radial striation. Slightly lirate inside from umbones to halfway with 8-9 elevated riblets. Small equal ears. Outside colour straw-yellow to light brown. Inside white. *Indo-Pacific.*

Note. This shell was used by Oyama, 1951 as type species for his genus *Flavamussium* regarded as synonymous with *Propeamussium*.

Propeamussium dalli (E.A. Smith, 1885)
pl. 27 fig. 4

Medium size, somewhat oblong, equivalve, thin and fragile, transparant, whitish, sometimes pale violet in colour, glossy. Right valve has fine concentric growth lines and fine radial substriae. The left valve is decorated with close concentric lirae with most minute crystal-like knobs. Ears small and equal. Inside glossy with approx. 10 fine radiating costellae extending only halfway each valve, they do not reach the margins. *Texas, West Indies, Bermuda.*

Note. This species is the largest of the *Propeamussiidae*. It was designated as type species for the genus *Paramussium* Verrill, 1897. Later placed in *Variamussium*. However, both of these are considered to be synonyms of *Parvamussium*.

Propeamussium diomedeum Dall, Bartsch & Rehder, 1938

Small shell (approx. 6.5 mm) translucent, inequivalve. Right valve fits into the left one. Right valve shows close, week concentric threads. Left valve decorated with numerous radiating threads and concentric lamellae. At the junctions small nodules. Interior reinforced by 10 radiating riblets. Colour whitish. *Hawaii.*

Propeamussium kauaium Dall, Bartsch & Rehder, 1938

Small shell (10 mm), equivalve, transparent. Right valve has a sculpture of concentric cords and numerous fine radiating lirations. Left valve has strong rounded hollow concentric cords, more concentrated towards the margins and a great number of radiating threads, giving the valve surface a cancellated sculpture. Interior of the valves have 10 strong radial ribs. Unequal ears. *Hawaiian Islands; dredged from deep water near Kauai.*

Propeamussium malpelonium (Dall, 1908)

Small to medium size, thin and fragile. Convex. Left valve with fine concentric lines and numerous fine radial lines. Right valve concentrically striated. Interior has 9 elevated ribs on each valve. Colour dirty white. Small subequal ears. *Recorded from Gulf of Panama, near W. Colombia and north of Marquesas Islands in deep water.*

Propeamussium meridionale (E.A. Smith, 1885)

Small (14 mm), thin and fragile white shell, somewhat oblique and gaping. Left valve rather convex, has numerous thin, radial lirae and concentric lines. The right valve flatter, has fine concentric grooves and ridges. The valves are internally supported by 11 ribs extending over the area from umbo to mid-body. Interior glossy and slightly iridescent. *Far south of Madagascar and far west of Central Chile in very deep water.*

Propeamussium molokaium Dall, Bartsch & Rehder, 1938

This is a tiny shell, the authors described only the right valve, since no left valve was found. *Hawaii, dredged near Molokai.*

Propeamussium nesiotum Dall, Bartsch & Rehder, 1938

Small shell (5 mm), whitish. Right valve has concentric laminae well raised near the ventral edges. Left valve has numerous radiating cords and concentric threads. At the meeting points weak nodules. Interior of the valves milky white and reinforced by 12 radiating riblets. *Hawaiian waters.*

Propeamussium pailoloum Dall, Bartsch & Rehder, 1938

Small shell (9 mm), flat, transparant, unequal ears. Right valve smooth in umbonal area but develops fine concentric threads increasing in size towards the ventral edges. As in right valve, top part of left valve smooth. Further down however concentric threads increase in strenght. In the lower two-thirds of the valve, they develop into thin and narrow wavy lamellae. Interior side of the valves supported by 10 radial riblets per valve. *Hawaiian waters.*

Propeamussium pourtalesianum (Dall, 1886)

Syn. *P. marmoratum* (Dall, 1886).

Small transparent and thin, slightly convex species. One valve with minute concentric grooves, other valve smooth. Interior of both valves carries 8 or 9 opaque riblets. Rather colourless. *West Indies and southern parts of Florida, from shallow to deep water.*

Propeamussium propinquum (E.A. Smith, 1885)

Small (approx. 7.5 mm), fragile, pellucid, orbicular, flat, glossy shell with small ears. Right valve smooth, left valve also, except for the numerous very fine hairline lirae. *Azores.*

Propeamussium sayanum (Dall, 1886)

Small, compressed, equivalve, thin shell. Right valve with about 30 tiny riblets, about 10 or 12 thereof, running right down to the margin, and crossed by concentric lamellae. Left valve has no riblets, the concentric sculpture is weaker. Unequal ears. Colour white. *West Indies, in deep water.*

Propeamussium scitulum (E.A. Smith, 1885)

Thin, tiny shell (4.5 mm,) compressed, semi-transparent and white. Right valve slightly deeper with concentric lines. Left valve has a minute cancellated appearance. Ears unequal. Valves carry 10 internal lirae. *New Guinea waters.*

Propeamussium sibogai (Dautzenberg & Bavay, 1904)
pl. 27 fig. 1, 1a

Small to medium size, to approx. 20 mm, thin, orbicular, translucent, slightly convex, semi-glossy shell. Exterior of the valves seems smooth but upper valve has numerous fine concentric and radial striae, forming a fine network. Left valve smoother. Each valve has 7 internal distincly raised rounded ribs which do not reach the margins. One valve slightly larger than the other. Very small equal ears. Horn-coloured, the ribs a strong orange-brown. *Recorded from Celebes and Bali, Indonesia and Japanese waters.*

Note. Type species of *Luteamussium* Oyama, 1951, which is considered a synonym of *Propeamussium*.

Propeamussium weberi (Dautzenberg & Bavay, 1912)

Oval, small, thin, translucent, smooth, slightly convex and gaping at the sides. Small equal ears. Upper valve bears very fine nearly invisible radial and concentric lines. The inside of the valves is decorated with 9 riblets starting from the umbonal area, but not reaching the ventral margins. Colour yellowish with small variation in shades between upper and lower valves. *Indonesian waters.*

Propeamussium watsoni (E.A. Smith, 1885)

Medium large, circular, thin, fragile shell, gaping both sides above centre. Right valve thinner, cream or light brown in colour. Left valve milky white and more opaque. Glossy and silky. Both valves finely concentrically lirate. The lirae towards the outer margins in the right valve narrowly lamellated. Left valve has also numerous radial lirae elevated at the umbo and then diminishing to only substriation. Equal ears. Interior glossy, strengthened with 5-6 pairs of white riblets. *Papua New Guinea.*

PARVAMUSSIUM Sacco, 1897

Type species: *Pecten duodecimlamellatum* Born, 1831 (fossil).

Differs from *Propeamussium* mainly by having a prominent byssal sinus and the valves being generally more convex and rarely gaping.

Parvamussium alaskense (Dall, 1871)
pl. 26 fig. 8

Syn. *P. simanense* Kuroda, 1931

Small shell (approx. 20 mm), inequivalve, suborbicular, translucent. Relatively wide, unequal, large ears. Lower valve depressed, smaller than upper one, bears many concentric ridges and has a soft epidermis with radial striae. Pinkish colour. Upper valve somewhat larger, more convex, with a granulose sculpture and numerous radial, irregular, often worn cords. Pinkish with some salmon colour on umbo. Inside of both valves 21 rounded radial white ribs. *Bering Sea to California, Japanese waters.*

Parvamussium cancellatum (E.A. Smith, 1885)

Small (20 mm), flattened, dull white in colour. Right valve deeper, with numerous fine radial lines crossed by concentric ones giving it a cancellate sculpture. Left valve has a silky appearance and has minute close concentric lirae. Small ears. Inside valves glossy and supported by 12 white, strong, radial riblets and some finer ones. *Florida, Bermuda, West Indies.*

Note. Assigned to *Variamussium* by Sacco (1897) and used as type species. Now considered a synonym of *Parvamussium.*

Parvamussium cmadoritinctum Kuroda, 1931.

Shell small, somewhat scaly radial ribbing on the left valve. Eight to nine internal ribs. Colour, angular patches and rays of orange. *Japan in deep water.*

Note. Placed in *Ctenamusium* by Iredale (1939). Considered to be a synonym of the genus *Parvamussium.*

Parvamussium cristatellum (Dautzenberg & Bavay, 1912)

This is a very small, whitish, rather flat shell with greatly unequal ears, and 9-10 interior ribs. Exterior with numerous concentric lines on lower valve and radial striae on upper valve. *Indonesian waters and Indian Ocean.*

Note. This species, named originally *Amussium cristatum* Bavay, 1905, has been renamed, since the latter name was already attributed to a Pliocene fossil by Bronn in 1828.

Parvamussium formosum (Melvill & Standen, 1907)

Small, whitish shell (approx. 9 mm), fragile, inequivalve, translucent. Left valve the larger, smooth. Right valve has minute concentric lines. Both valves have 10-12 internal ribs, which on the left valve do not reach the ventral margin. Unequal ears. *Red Sea, Persian Gulf.*

Note. Non *Amusium japonicum formosum* Habe, 1964.

Parvamussium holmesii (Dall, 1886)

Small (up to 25 mm), thin, orbicular, unequal valves. The convex one has about 20-28 radial striae weakening towards the umbo. Interspaces have concentric lines which are elevated as small nodules where they meet the radials. The other valve, about flat, has only wavy concentric grooves. Ears unequal. Colour yellowish white often with brownish flecks. Inside has about 12 pairs of radial striae or riblets which do not extend to the ventral edges. *Florida and Caribbean in deep water.*

Parvamussium hoskynsi (Forbes, 1844)

Small shell (approx. 9 mm), upper valve covered with radial rows of tiny frills. Lower valve closely concentrically striated. About 14-18 fine interior riblets per valve. Colour white, orangy-yellow inside toward umbonal area. *Mediterranean, East Atlantic from Greenland and Norway down to the Azores.*

Note. Type species for the genus *Xenamussium* described by Oyama (1944) and now considered a synonym of *Parvamussium.*

Parvamussium ina (Dautzenberg & Bavay, 1912)

Small, fragile, translucent, glossy, orbicular, equivalve, slightly convex species. Ears unequal. Externally sculptured with numerous and hardly visible concentric striae. Interior of the valves supported by 8 radial riblets. Straw-coloured with some reddish-brown, inside same colour but paler. *Indonesia, in deep water.*

Note. Very close to *P. torresi* which shell however carries 12 internal riblets.

Parvamussium intuscostatum (Yokoyama, 1920)

Small shell. Left valve imbricated, right valve has concentric grooves only. About 15 internal riblets. Tinted with orange-red, pinkish and white. *Japan, in deep water.*

Note. Named *Polyneamussium intuscostatum* by Habe in 1951 and used as type species for this genus. Now treated as a synonym of *Parvamussium.* A subspecies called *P. intuscostatum sawanensis* (Hertlein, 1931) replaces *P. intuscostatum multicostatum* (Yokoyama, 1926).

Parvamussium jeffreysii (E.A. Smith, 1885)
pl. 27 fig. 2

Small, slightly elongated compressed and somewhat oblique, gaping at the sides. Valves very thin. Right valve, deeper, is the more brightly coloured, upper and central portion bright orange, outer border paler and transparent. Sculptured with numerous fine round lirae radiating from the umbo and extending to the margins all around. Also marked with fine concentric lines together producing a distinct cancellated surface. Left valve has numerous concentric lines only. Small ears. Interior very glossy with 15 white radial lirae terminating abruptly and extending over the tinted part of the valves. *Japan, Philippines, Indonesia, Indian Ocean.*

Note. It is the type species of *Bathyamussium,* a synonym of *Parvamussium.*

Parvamussium lacteum (Dautzenberg & Bavay, 1912)

Tiny (6 mm) but solid, orbicular shell, slightly convex, equivalve. Smooth ears. Upper valve carries numerous shallow fine radial riblets and concentric grooves resulting in a trellis-like sculpture. Lower valve has no ribs but fine grooves only. The interior of both valves counts 8-10 radial riblets stronger near the margins and disappearing near the umbones. *Moluccas, in deep water.*

Parvamussium lucidum (Jeffreys in Thomson, 1873)

Small shell (10 mm), roundish, sometimes oval, flattened, thin, semi-transparent, glossy. Upper valve has no sculpture, only fine concentric close growth lines. Lower valve smaller, marked with numerous concentric striae. Small sub-equal ears. Colour white. Inside pearly with 9-11 slight, thread-like riblets visible through the valve and occupying the middle of the interior, terminating in tiny rounded knobs. *Atlantic around Azores.*

P. lucidum forma *striata* (Jeffreys, 1876).
Upper valve marked by fine close-set distinct longitudinal striae. With 9-15 interior ribs.

Parvamussium maorium (Dell, 1956)

Small, thin, almost circular, slightly convex, transparent. On the inside the valves bear 10 raised radial ribs which do not reach the ventral margins. They are visible through the valves and are seen as white on the left and as orange on the right valve. Left valve covered with numerous fine concentric lines. Right valve has some striation near the ventral margin. Subequal ears. *South Island new Zealand, deep water.*

Parvamussium margaritiferum (Dautzenberg & Bavay, 1912)

Very small, thin, somewhat elongated shell with small near-equal ears. Upper valve not unlike a *Chlamys*, with 25 external riblets and many other small ones in between. Fine concentric lamellae cross the riblets and form small pearly nodules, hence the name. Lower valve smooth at umbo but further with small grooves. The valves have 11 internal ribs, not reaching the margins and visible from the outside through the lower valve only. Colour upper valve light yellow with some soft rose. Lower valve lighter in colour. *Moluccas, in deep water.*

Parvamussium octodecimliratum (Melvill & Standen, 1907)

Small, nearly equilateral, paperthin shell with unequal ears. Right valve carries fine cancellated striae. Left valve has concentric grooves. Interior of the valves bear 18 strong lirae each. *Antarctica, in very deep water.*

Parvamussium pauciliratum (E.A. Smith, 1903)

Tiny, light, smooth, semi-transparent, slightly convex, orbicular shell with subequal ears. The flatter of the two valves is straw-coloured with brown streaks and reddish dots radially arranged. The more convex valve is transparent and colourless. *West and South Pacific, in deep water.*

Parvamussium rubrotinctum (Oyama, 1951)

Small size, thin, flat, oval shaped shell with subequal valves. Right valve has vague concentric ribbing and about 12 internal lirae. Left valve generally smooth with some indistinct concentric undulations and about 12-18 internal lirae. Colour right valve pale orange with white internal lirae showing through as rays, left valve creamy with orangy tints, lirae showing as white lines. *Southern Japan.*

Parvamussium salacon (Iredale, 1929)

Small shell. Right valve closely concentrically ridged. Left valve has radial ridges also, sculpture becoming cancellate when ageing. Ribbing inside. Colour whitish and translucent. *Australia.*

Note. This species was placed in the genus *Ctenamusium* Iredale, 1939, which is a synonym of the genus *Parvamussium.* Not unlike *C. thetidis,* but stronger in sculpture and different in internal ribbing.

Parvamussium siebenrocki Sturany, 1901

Related to *P. formosum,* but not so smooth. The internal ribbing extends to the ventral margins. Unequal ears. *Red Sea, Gulf of Oman.*

Parvamussium siratama (Oyama, 1951)

Small (7.5 mm), almost circular thin shell, nearly equilateral and equivalve. Valves seem smooth to the eye, although left valve has minute radial striations. Small ears. Left valve whitish, right valve fawn-coloured. Inside glossy with 10 lirae per valve. *Japan, in deep water.*

Note. Type species of *Micramussium,* which is a synonym of *Parvamussium.*

Parvamussium squamigerum (E.A. Smith, 1885)

Small, slightly convex, rounded, thin, pellucid, whitish. Right valve has numerous fine concentric lines and 15-30 radial ridges. Concentric lamellae elevated to form scales. Small, almost equal ears. Inside shows 10-11 lirae per valve. *West Indies, Bermuda.*

Note. Used by Oyama, 1952 as type species for his *Squamamussium,* which is synonymous with *Parvamussium.*

Parvamussium texturatum (Dautzenberg & Bavay, 1912)

Very tiny, thin, orbicular shell with relatively large and unequal ears. Upper valve cancellated, composed of some 17 radial riblets and smaller ones in interstices, crossed by concentrically arranged scales at regular distances. Lower valve has numerous fine grooves only. Interior of each valve shows 16 riblets not reaching the margins. Straw-coloured with some reddish blotches on upper valve and white on lower valve. *Sulu Archipelago.*

Note. According to K. Bernard (1964) also found in quantities off the *Natal Coast, S. Africa.*

Parvamussium thalassinum (Dall, 1886)

Syn. *P. fenestratus* Verrill, 1881 (non Forbes, 1844)
P. actoni (Martens, 1853)
P. philippe (Anton, 1855)

Shell small with very unequal ears. Upper valve cancellated. Some internal ribbing. Colour upper valve reddish with creamy mottling, lower valve white. *West Indies and East coast USA.*

Parvamussium thetidis (Hedley, 1902)

Small (6 mm), thin, flattened, white and translucent shell. Valves externally glossy with concentric grooves. Anterior ears well developed. Long hinge line. Inside each valve shows 10 slender raised radial lirae knobbed at the ventral margins and disappearing near the umbones. Lirae visible through the valves. *S. Queensland to Tasman Sea, E. Australia, also reported from SW. Australia.*

Note. Used by Iredale, 1939 as type species for his *Ctenamusium*, which is considered to be a synonym of *Parvamussium*.

Parvamussium torresi (E.A. Smith, 1885)

Small shell (8 mm), orbicular, very thin, right valve smooth and glossy, left valve carries numerous concentric lines. Valves have 12 internal radial lirae. Colour white to cream. *N. Queensland, Australia, in deep water.*

Parvamussium zoniferum (Dautzenberg & Bavay, 1912)

Tiny, thin, orbicular, translucent, slightly convex. Auricles small. Lower valve shows a great number of tiny regular concentric grooves. Upper valve rather smooth. Both valves carry 11 interior riblets which do not reach the margins. Consequently the edges have little support and are very brittle. Colour white. *Indonesian waters.*

CYCLOPECTEN Verrill, 1897

Type species: *Pecten pustulosus* Verrill, 1873.

Shell small, orbicular but somewhat oblique, thin, convex, inequivalve. Not gaping. Most species have opaque, some others however have translucent valves. Exterior sculpture of right valve lamellose, or concentric ridges. Right valve, upturned at the margin, and the smaller of the two valves, fits in the left (or upper) valve. The latter scaly or reticulated. Sometimes both valves smooth. Valve edges regular without folds, ribbing or crenulations. No internal ribbing. Distinct byssal notch.

subgenus CYCLOPECTEN Verrill, 1897

Cyclopecten (Cyclopecten) acutus Grau, 1959

Very small (3-4 mm), thin, opaque, moderately convex, orbicular. Both valves covered by thin layer of transparent material, extending beyond entire margin of the valve. Smooth except for some weak concentric ridges. Ears unequal. Colour white to yellow white. In addition, left valve has some streaks or spots of yellow to brown. *West Colombia.*

Cyclopecten (Cyclopecten) aupouria Powell, 1937

Very small, fragile white shell. The left valve has prominent concentric lamellae and radial threads in interstices. Only a few valves have been found, so far no complete shells. *New Zealand.*

Cyclopecten (Cyclopecten) aviculoides (Smith, 1885)

Minute shell (2 mm), dirty white in colour, thin, oblique and only slightly convex. Right valve smaller and fits into the somewhat larger left valve. The latter bears relatively coars lamellae and 7-8 strong radial riblets or threads, together resulting in a cancellated surface. Equal ears. *Prince Edward Island, Canada in deep water.*

Note. The shape of the shell remotely like the bivalve genus *Avicula.*

Cyclopecten (Cyclopecten) barbarensis Grau, 1959

Tiny, thin shell (4 mm), valves translucent. Left valve considerably convex, carrying up to 32 rows of concentric lamellae and up to 32 radial ridges, resulting in a trellis-like sculpture. Right valve only slightly convex with a great number of fine concentric ridges. Ears unequal. *Southern California, near Santa Barbara Island.*

Cyclopecten (Cyclopecten) benthalis Grau, 1959

Small (approx. 8 mm), translucent, somewhat oblong, equivalve. Left valve convex with 13 radial main ridges, having tiny nodules at regular intervals and some additional ridges in between. Right valve flatter with numerous concentric ridges. Unequal ears. *California, S. Nicolas Island.*

Cyclopecten (Cyclopecten) catalinensis (Willett, 1931)

Small (7 mm), thin, transparent. Left valve shows fine growth lines and microscopic striae near the umbones. Right valve has concentric fine grooves from halfway the valve down to the margins which are reflexed in the ventral area. *Southern California to Mexico, Gulf of California.*

Cyclopecten (Cyclopecten) clathratus (Martens, 1881)

Very small (approx. 3 mm), very inequivalve, thin, glossy, translucent and rests in the left valve, it bears minute growth lines and a hardly visible radial somewhat granular sculpture. The left valve is deeper, bearing a sculpture of fine concentric lamellae and very fine radial ones, resulting in a minute network. Ears almost equal. *Kerguelen Island, near Antarctica.*

Cyclopecten (Cyclopecten) cocosensis (Dall, 1908)

Small (9 mm), rather orbicular, translucent white shell, decorated with irregular lines and blotches in reddish brown and white. Left valve smooth with vague radial striae, right valve shows numerous fine concentric lines. Posterior part of the shell has a distinct concave fold or furrow. Ears small and subequal. *Costa Rica, Cocos Island.*

Cyclopecten (Cyclopecten) eous (Melvill & Standen, 1907)

Small shell, thin, translucent, near orbicular. Right valve the more convex, both valves smooth but for some vague concentric striae on the left valve. Ears sub-equal. Colour off-white. *Persian Gulf, in deep water.*

Cyclopecten (Cyclopecten) exquisitus Grau, 1959

Tiny shell (3.5 mm), opaque, flattened, thin. Unequal ears. Left valve slightly more convex than right valve, covered with a great number (up to 40) of rows of small concentric lamellae. Right valve very flat with numerous radial ridges, overlapped by concentric rows of small lamellae. Colour white or yellowish. Left valve sometimes with streaks of yellow, red-brown, orangy-red or brown. *Northern Gulf of California to Peru. Also recorded from the Galapagos Islands.*

Cyclopecten (Cyclopecten) favus Hedley, 1902

Syn. *C. nepeanensis* Pritchard & Gatliff, 1904

Small, approx. 3 mm, thin shell. Valves dissimilar in shape, size and sculpture. Left valve larger and the more convex one, transparent at the margins, smooth except for a few fine concentric growth lines; ventral margin rounded. Right valve smaller and overlapped by the other, less oblique, sculptured by concentric growth lines and superficially by numerous fine irregular radiating threads, connected with close-wet transverse bars resulting in honeycomb structure which, however, peels off in places, 12 riblets on inside of each valve. Colour opaque pale yellow. Near equal ears. *South and South-West Australia, including Tasmania.*

Cyclopecten (Cyclopecten) gaussianus (Thiele, 1912)

Small rounded shell (3.5 mm), very thin, translucent, with a great number of concentrical ridges and grooves on the valves. Distinct ears. *Antarctica.*

Cyclopecten (Cyclopecten) hexagonalis Powell, 1958

Small, rare shell from *Antarctica.*

Cyclopecten (Cyclopecten) imbrifer (Lovén, 1846)

Small (10-11 mm), flat, thin, rounded, unequal ears. Right valve, carrying minute concentric striae, is the smaller and rests within the left valve which has very fine concentric lines and irregular radial arrangements of tiny nodules. Inside smooth. Colour white to brownish grey. *Arctic Seas and Northern Atlantic, in deep water.*

Cyclopecten (Cyclopecten) incubans Barnard, 1964

Small (3.5 mm), thin, translucent, white, suborbicular shell with distinct ears. Right valve, smaller than left one, has minute close concentric striae, ribs very obscure. Left valve has concentric striae and very fine and close radial ribbing (up to 50). Interior smooth. *South Africa, in deep water.*

Cyclopecten (Cyclopecten) kermadecensis (E.A. Smith, 1885)

Small (7.5 mm), very thin, fragile, orbicular, white, very flattened, transparent shell. Right valve shows close concentrical lirae. Left valve has numerous vague concentrical lirae with many radial series of small scales. Unequal ears. *Kermadec Islands, in deep water.*

Cyclopecten (Cyclopecten) leptaleus Verrill, 1884

Small (7 mm), thin, fragile, orbicular shell. Upper valve has fine concentric grooves, becoming obsolete near the umbones. Lower valve, which is the flatter, also carries minute concentric lines. Whitish translucent. Ears slightly unequal. *Arctic and Northern Atlantic, in deep water. Also reported from Carolina and Florida.*

Cyclopecten (Cyclopecten) liriope (Dall, 1908)

Small, near circular, thin shell. Left valve convex with numerous fine radial cords and fine concentric sculpture. Right valve the same but radial sculpture not too obvious, concentric lamellae more distinct. Outside colour off-white, inside glass-like with the outside sculpture visible through the transparent valves. Ears small and unequal. *Gulf of Panama and the Galapagos Islands.*

Cyclopecten (Cyclopecten) mestayerae Dell, 1956

Tiny, fragile, rounded shell of a white colour. Sculpture described from a left valve having "a few minute radial striations". *New Zealand, North Cape, in deep water.*

Cyclopecten (Cyclopecten) nanus Verrill & Bush, 1897

Small (7 mm), nearly round, thin, translucent, only slightly convex. Large and broad ears. Valves bear thin concentric striae and sometimes minute radial cords. No internal lirae. Diffused milky colour with light brown or reddish brown spots on right valve mainly. *West Atlantic, USA to Brazil, in deep water.*

Cyclopecten (Cyclopecten) obliquus Hedley, 1902

Resembles *C. favus* but is not that small (5 mm). Flattish, very thin, semi-transparent, white. Under magnification the hexagonal spaces appear to be stippled. This shell is dull, not glossy and has 16 interior radial ridges. The left valve is readily recognized by its elevated concentric lamellae, about 35 in number. Ears very unequal. *Australia, Tasman Sea.*

Cyclopecten (Cyclopecten) perplexus Soot-Ryen, 1960

Tiny shell (approx 1.5 mm) semi-translucent, slightly oblique and convex, inequivalve. Colour white. Left valve more convex than right one, larger umbones, has varying number of threadlike radial and concentric lines carrying thin toothlike hollow projections. Right valve, less convex, has fine concentric lines and indistinct radial sculpture. Unequal ears. *South Atlantic.*

Note. Species very similar in shape to *Cyclopecten transenna* as well as *Cyclopecten barbarensis*. A noteworthy fact about this species is that the larvae are retained in the gills of the animal and according to Soot-Ryen the first record of broodprotection in the pectinids.

Cyclopecten (Cyclopecten) powelli Dell, 1956

Fragile, very small, rounded shell with unequal ears. Described from a left valve which has faint minute concentric raised threads, crossed by a few radial lines. Apparently close to *Cyclopecten aupouria. Tasman Sea and New Zealand.*

Cyclopecten (Cyclopecten) pteriola (Melvill & Standen, 1907)

Shell small (4 mm), inequivalve, somewhat oblique. Right valve, the flatter, has rough concentric striae. Left valve, more convex, has fine concentric cords. Unequal ears. Grey in colour, dull. *South Atlantic.*

Note. Somewhat similar to the minute *C. aviculoides* but different in sculpture.

Cyclopecten (Cyclopecten) pustulosus (Verrill, 1873)

Thin, small shell (approx. 10 mm), orbicular, unequal ears and usually translucent. Lower valve, the flatter, has narrow raised concentric lamellae, sometimes has imbrications. The upper, more convex valve, has about 14 radial rows of tiny knobs resembling blisters, which when worn or damaged, form tiny scales, and further has concentric raised lines. No internal ribbing. Pale yellow-white colour. *Central East coast of USA, in deep water.*

Note. This shell has been referred to *Propeamussium hoskynsi* by Jeffreys, and to *Cyclopecten imbrifer* by Dall. However, it does not possess internal ribs like *hoskynsi*, although it has similar valve decorations. From *imbrifer* it differs in the sculpture of the left valve.

Cyclopecten (Cyclopecten) reticulus (Dall, 1886)

Small (7 mm), thin, somewhat oblique, rounded, inequivalve. The right valve is larger and more convex and carries minute radial grooves and concentric lines, resulting in a fine sharp sculpture where they meet. Left valve smaller and flatter with concentric lines only. Whitish colour. *Gulf of Mexico and Central West Atlantic, in deep water.*

Cyclopecten (Cyclopecten) rotundus (Dall, 1908)

Syn. *P.(C.) pernomus* Hertlein, 1935

Very small (3 mm), thin, roundish, white shell. Valves almost equally convex, but right valve slightly smaller than the left one, unless under magnifying glass surface seems smooth and polished. Left valve has fine radial striae. Ears subequal. White variegated with purplish brown, sometimes yellowish-white only. *California to Ecuador.*

Cyclopecten (Cyclopecten) secundus Finlay, 1927

Syn. *P. hinemoa* Mestayer, 1927

Small, flattish, rounded, translucent, equilateral shell with distinct latice-like sculpture. Whitish colour. *New Zealand, North Island.*

Note. C. secundus is larger and the sculpture more pronounced than in *C. transenna.*

Cyclopecten (Cyclopecten) simplex Verrill, 1897

Tiny shell (4 mm), thin, flattened, transparent, rounded. Unequal ears. Left valve almost smooth, right valve is covered with numerous fine concentric grooves. *West Indies.*

Cyclopecten (Cyclopecten) strigillatus (Dall, 1889)

Small shell (10 mm), with very unequal ears, moderately convex. Both valves covered with concentric lamellae. Colour white. *Florida and Caribbean Sea.*

Note. There is quite some confusion about the taxonomy of this species. Some assign it to *Palliolum*, others to *Pseudamussium.*

Cyclopecten (Cyclopecten) subhyalinus (E.A. Smith, 1885)

Small (7 mm), flattened, translucent and shiny. Left valve medium convex with some concentric and radial sculpture. Right valve has more distinct concentric striae. Unequal ears. *Southern coast of Chile, in deep water.*

Cyclopecten (Cyclopecten) subimbrifer Verrill & Bush, 1897

Small shell, inequivalve, orbicular and translucent. Unequal ears. Left valve is covered with somewhat elevated concentric lirae interrupted by an arrangement of about 40 radial series of small arched scales. Right valve smaller than the left and covered with fine concentric raised lines. Colour dirty white. *West Atlantic.*

Cyclopecten (Cyclopecten) thielei Powell, 1958

Rare, small shell *off the Antarctic coasts.*

Cyclopecten (Cyclopecten) transenna (Suter, 1913)

Tiny, fragile, translucent and orbicular, white. Left valve with open lattice sculpture formed by radial ribs crossed by concentric lamellae and at the meeting-points tiny knobs. Right valve smaller and flatter. Sculpture concentrically and radially only weak. Very unequal ears. *New Zealand, South Island.*

Cyclopecten (Cyclopecten) vidalensis Barnard, 1964

Described from only one left valve of *Zululand, coast of South Africa.* Provisionally placed in *Cyclopecten,* but, according to Bernard "if this valve is juvenile, older shells might show better developed internal costae, which would indicate removal into another genus".

Cyclopecten (Cyclopecten) zephyrus Grau, 1959

Small, very thin and fragile, translucent shell (approx. 8 mm). Left valve rather convex with fine but few concentric ridges and some radial cords, not always reaching the margins. Right valve less convex, reflexed margins, numerous concentric ridges. *California.*

subgenus DELECTOPECTEN Stewart, 1930

Type species: *Pecten vancouverensis* Whiteaves, 1893.

Cyclopecten (Delectopecten) arces (Dall, 1913)

Medium sized shell (35 mm), transparent, white, suborbicular, slightly convex, one valve more so than the other, entire surface of the valves decorated with radial treads and similar concentric threads, forming almost equal reticulation at the intersections, more or less spinose towards the valve margins. No distinct anterior ear, being one with the valve. Inside glossy. *Off Santa Barbara and Santa Cruz Island, California.*

Note. Considered to be identical with *C.(D.) randolphii tillamookensis* (Arnold, 1906).

Cyclopecten (Delectopecten) bistriatus (Dall, 1916)

Small, thin, nearly round. Both valves slightly convex. Right valve has fine concentric lamellose sculpture, left valve ditto, but has also numerous fine radial raised cords. Colour white. *South California, Lower California, in deep water.*

Cyclopecten (Delectopecten) culebrensis (E.A. Smith, 1885)

Very small (8 mm), thin, flattened and translucent, somewhat elongated. Left valve has rough, thin ribs and concentric striae and lamellae. Right valve, the flatter, has no ribs only concentric sculpture. Large anterior and small posterior ear. *West Indies and Azores.*

Cyclopecten (Delectopecten) distinctus (E.A. Smith, 1885)

Small (approx. 8 mm), inequivalve, compressed, thin, translucent, not glossy but rather dull, white in colour. Right valve has finer concentric lirae than the left valve and no radial lines. Left valve more convex, concentric lines and weak radiating lines. *Off Marion Island, Southern Indian Ocean, in deep water.*

Cyclopecten (Delectopecten) fosterianus (Powell, 1933)

Small, thin, equivalve, translucent shell. Valves without any colour, surface almost smooth, sometimes with some minute hollow nodulous scales and occasionally growth lines. One defined ear, the other more a continuation of the valve. *New Zealand, North Island, in deep water.*

Cyclopecten (Delectopecten) greenlandicus (Sowerby II, 1842)
pl. 27 fig. 7

Syn. *Pecten vitreus* Gray, 1824
Pseudamussium andersoni Dall, 1919
Pecten binominatus Hanna, 1924

Small (15 mm), orbicular, sub-equivalve, rather flat, thin, nearly smooth, transparent. Only left valve has a fine sculpture of thin, raised wavy cords, sometimes with additional fine striae and growth lines. Subequal ears. *Arctic, East Atlantic south to N.W. Africa, West Atlantic, Greenland.*

Note. Thiele (1935) used this shell as type species for *Arctinula,* a synonym of *Cyclopecten.*

Cyclopecten (Delectopecten) incongruus (Dall, 1916)

Small but not minute (14 mm), rounded, white shell. Left valve has beautiful sculpture consisting of many concentric sharp lamellae, crossed by radial, slightly elevated lines, tiny nodules at the meeting points. Right valve less ornate with numerous concentric grooves. *Off Cedros Island, Lower California, from deep water.*

Cyclopecten (Delectopecten) micaceus Dautzenberg & Bavay, 1912

Tiny, extremely thin and fragile, transparent, semi-orbicular and oblique shell, almost flat, equivalve. Appearance like a thin sheet of mica. Ears unusually large, not a separate feature off the shells as normally is the case, but united with and spread over the full width of the valves. Upper valve smooth with only concentric undulations, lower valve more or less the same but bears numerous very fine radial lines. *Celebes (Sulawesi). Dredged at a depth of 1300 metres near Makassar.*

Cyclopecten (Delectopecten) murrayi (E.A. Smith, 1885)

Small (15 mm), semi-orbicular, very flat, translucent, inequivalve shell. Valves have different sculpture, concentric lirae of the deeper valve are stronger than those on the right valve. Ears about equal. Inside pearly. Colour a dirty white. *N. Australia, very deep water, Cape York.*

Note. Type species of the synonymous subgenus *Catillopecten.*

Cyclopecten (Delectopecten) polyleptus (Dall, 1908)

Small (9 mm), thin, translucent, flattened species. Valves with numerous radial rows of nodulous scales, and concentric ridges. Ears equal. Colour white. Inside glossy. *Galapagos Islands and Southern Chile.*

Cyclopecten (Delectopecten) randolphi (Dall, 1897)

Syn. *P. whiteavesi* Orcutt, 1915

Small, but one of the relatively larger Delectopectens (approx. 27 mm), thin, generally smooth, sometimes with vague concentric undulations, rather flat orbicular valves. White, glossy. Very unequal ears. *Japan, Bering Sea to Mexico, in deep water.*

Cyclopecten (D.) randolphi tillamookensis (Arnold, 1906)

Shape as the nominal subspecies, but instead of smooth valves it has strong radial spinose and a number of concentric ridges, resulting in a cancellated sculpture. *Bering Sea to Mexico.*

Note. Some malacologists consider this subspecies identical to *Cyclopecten (D.) arces.*

Cyclopecten (Delectopecten) translucens Dautz. & Bavay, 1912

Tiny, fragile, transparent, orbicular shell, very flat, gaping at both sides. Ears unequal. Valves lirate and non-glossy with some very vague rays. Colour uniformly white. *Moluccas.*

Cyclopecten (Delectopecten) vancouverensis (Whiteaves, 1893)

Very small, thin equivalve, convex, translucent shell. Sculpture consists of numerous hardly visible fine, irregular, radial, raised lines. Unequal ears. Honey coloured. *From Bering Sea to Lower California, in deep water.*

Cyclopecten (Delectopecten) vitreus (Gmelin, 1791)
pl. 27 fig. 3

Syn. *Pecten abyssorum* Lovén in Sars, 1878
Pecten gelatinosus Mabille & Rochebrune, 1889

Small, rounded, slightly oblique, thin and translucent. Has only one distinct ear, the other ear forms part of the main body of the valves. Convex valves without ribs but with fine radial grooves and with fine concentric growth lines and minute knobs. White to golden in colour. *Arctic, Atlantic, Mediterranean and Indo-Pacific.*

Note. Locard named many varieties, such as *elongatus, denudatus, inflatus, minor,* etc.

C.(D.) vitreus abyssorum (Lovén in Sars, 1878) is more oblique, has no radial cords and therefore is smoother than the typical, with smaller ears. *Very deep water of more northern locations.*

C.(D.) vitreus chaperi (Dautzenberg & Fischer, 1897) is smaller and is opaque. Greyish white in colour.

C.(D.) vitreus laevigatus (Lovén in Sars, 1878) has no knobs or lamellae and is therefore completely smooth.

C.(D.) vitreus macrocheiricola (Habe, 1951) is a local *Japanese* subspecies (**pl. 27 fig. 8**).

C.(D.) vitreus gelatinosus is described *from S. Chile and Tierra del Fuego.*

Cyclopecten (Delectopecten) zacae (Hertlein, 1935)

Syn. *P. panamensis* Dall, 1908 (not Dall, 1898)

Small shell, very thin, flat and oblique, brittle and translucent. Valves have from 40-60 fine irregular radial striae of a scaly nature, crossed by fine concentric lines. Small unequal ears. Cream colour. *Lower California to Panama, in very deep water.*

subgenus HEMIPECTEN Adams & Reeve, 1849

Type species: *Hemipecten forbesianus* Adams & Reeve, 1849.

Cyclopecten (Hemipecten) forbesianus (Adams & Reeve, 1849)

Thin, transparent, small, orbicular shell. Upper valve convex with concentric lines. Lower valve smooth and flat. Colourless or rayed with orange. Ears very unequal with deep byssal notch. This species was described from 2 specimens *dredged in the Sulu Sea (W. Pacific).* Melvill (1906) reports it from *Bombay (India), the Laccadives and Maldive Islands.*

subgenus HYALOPECTEN Verrill, 1897

Type species: *Pecten undatus* Verrill, 1885 (not Defrance, 1825).

Cyclopecten (Hyalopecten) dilectus (Verrill & Bush, 1897)

Small, thin, fragile, translucent, slightly oblique, unequal ears. Surface of both valves covered with broad and regular undulations, most prominent in the left valve and crossed by regular thin raised lines, which are nearly obsolete on the right valve. Interior of the valves strongly undulated and left valve marked by distinct grooves. Colour off-white or yellow. *West Atlantic coast of USA, deep water.*

Cyclopecten (Hyalopecten) fluctuatus (Bavay, 1904)

Small (10 mm) translucent, slightly convex shell. Upper valve is decorated with about ten concentric minute folds and numerous small radial scale striae. On the lower valve also about ten concentric undulations, regularly distributed. Interior surface of both valves smooth. Unequal ears. Distinct byssal groove. Colour light-brown.

Note. This species is very similar to *C.(H.) frigidus* but is kept separate by Bavay as a *tropical shell*, wheras *C. frigidus* is found in colder waters.

Cyclopecten (Hyalopecten) frigidus (Jensen, 1903)
pl. 27 fig. 6

Syn. *P. fragilus* Jeffreys, 1876

Small, thin shell, slightly elongated and irregularly roundish, flattened, right valve more so than left. Translucent and silvery. Valves have approx. 12 concentric folds with numerous fine radiating striae. Small unequal ears. Inside glossy. *N. Atlantic, Spitsbergen to Iceland and Shetland Islands, in very deep water.*

Cyclopecten (Hyalopecten) neoceanicus (Dall, 1908)

Small (12 mm), thin, oblique, equivalve. Both valves somewhat undulated and reticulated by radial and concentric sculpture, but not nodulous. Small ears. Dirty white colour. *Galapagos Islands and Gulf of Panama, in very deep water.*

Cyclopecten (Hyalopecten) pudicus (E.A. Smith, 1885)

Small, but not minute (approx. 19 mm), whitish. Right valve flatter than left one. Resembles *C. fragilis* (Jeffreys) but differs somewhat in shape and having both valves plicate and lirate. Unequal ears. *Prince Edward Islands, South Indian Ocean.*

Note. Closely related to *C.(H.) dilectus.*

Cyclopecten (Hyalopecten) undatus (Verrill & S. Smith, 1885)

Small, (19 mm), thin, fragile, oblique, transparent, nearly equivalve. Valve carry numerous fine regular riblets and show growth lines. Unequal ears. Colour whitish. *E. Atlantic (Europe to W. Africa), W. Atlantic (east coast USA) and from Antartic.*

subgenus SIMILIPECTEN Winckworth, 1932

Type species: *Pecten similis* Laskey, 1811.

Cyclopecten (Similipecten) similis (Laskey, 1811)

Syn. *P. tumidus* Turton, 1819

Very small, thin, fragile, circular. No ribbing. Subequal ears. Left valve larger than right one. Sculpture consists of fine concentric and radiating lines. Colour cream or pale yellow, with white or brownish mottling. *Atlantic from Norway to West Africa, and the Mediterranean.*

LITERATURE

ABBOTT, R.T., 1954. American seashells: 1-541. New York

—, 1962. Sea shells of the world: 1-160. New York.

ADAM, W., 1960. A propos de Chlamys abscondita (P. Fischer, 1898) de la côte occidentale de l'Afrique. Bull. Inst. r. Sci. nat. Belg. 36(20):1-10.

ADAMS, A. & L. REEVE, 1848. The zoology of the voyage of the H.M.S. Samarang; Mollusca: 1-87. London.

ADAMS, H. & A. ADAMS, 1858. The genera of Recent Mollusca 2:661.

ALLAN, J., 1962. Australian shells: 1-487. Melbourne.

BARNARD, K.H., 1964. Contributions to the knowledge of South African marine Mollusca. Part V. Lamellibranchiata. Ann.S.Afr.Mus. 47: 361-593.

BAVAY, A., 1903. Note sur quelques espèces du genre Pecten, nouvelles ou mal connues. J. Conch. Paris 50: 399-406.

—, 1904. Descriptions de quelques nouvelles espèces du genre Pecten et rectifications. J. Conch. Paris 52: 197-206.

—, 1904. Espèces nouvelles du genre Pecten provenant de "L'Indian Museum de Calcutta". Mém. Soc. zool. Fr. 17: 186-190.

—, 1905. Sur quelques espèces nouvelles, mal connues ou faisant double emploi dans le genre Pecten. J. Conch. Paris 53: 18-30.

—, 1905. Sur quelques espèces ou varietés nouvelles du genre Pecten. J. Conch. Paris 53: 243-247.

BERNARDI, A., 1858. Description d'espèces nouvelles. J. Conch. Paris 7: 90-92.

BERRY, S.S., 1968. Notices of new eastern Pacific Mollusca-VII. Leafl. Malac. 1(25): 155-158.

BLOMMERS, J. et al., 1975. De Europese Pectinidae. Marien Biologische Publicatie 2: 1-64.

CAMERON, R., 1967. Shells: 1-128. London.

CHENU, J.C., 1859. Manuel de conchyliologie et de paleontologie conchyliologique 2. Paris.

CLARKE, A.H., 1965. The scallop super-species Aequipecten irradians (Lamarck). Malacologia 2: 161-188.

COOMANS, H.E., 1973. Pearl formation in Chlamys varia from Bretagne. Basteria 37: 141-143.

COTTON, B.C., 1960. A new species of scallop Notochlamys hallae sp.nov. from South Australia. Malac. Sect. R. Soc. S. Aust. 14: 1-2.

—, 1961. South Australian Mollusca. Pelecypoda: 1-363. Adelaide.

DALL, W.H., 1871. Descriptions of sixty new forms of mollusks from the west coast of North America and the North Pacific Ocean, with notes on others already described. Am. J. Conch. 7: 93-160.

—, 1897. New west American shells. Nautilus 11: 85-86.

—, 1902. Illustrations and descriptions of new, unfigured, or imperfectly known shells, chiefly American, in the U.S. National Museum. Proc. U.S. National Mus. 24: 499-566.

—, 1908. Reports on the dredging operations off the west coast of central America to the Galapagos, to the west coast of Mexico, and in the Gulf of California, in charge of

Alexander Agassiz, carried on by the U.S. fish commission steamer "Albatross" during 1891, Lieut. commander Z.L. Tanner, U.S.N., commanding, XXXVIII., and Reports on the scientific results of the expedition to the eastern tropical Pacific in charge of Alexander Agassiz, by the U.S. fish commission steamer "Albatross", from October, 1904, to March 1905. Lieut. commander L.M. Garrett, U.S.N., commanding, XIV., Reports on the Mollusca and Brachiopoda. Bull. Mus. Comp. Zool. 43: 203-487.

—, 1909. Report on a collection of shells from Peru, with a summary of the littoral marine Mollusca of the Peruvian zoological province. Proc. U.S. Nat. Mus. 37: 147-294.

—, 1913. Diagnoses of new shells from the Pacific Ocean. Proc. U.S. Nat. Mus. 45: 587-597.

—, 1914. Notes on some West American pectens. Nautilus 27: 121-122.

—, 1921. Summary of the marine shellbearing mollusks of the Northwest coast of America, from San Diego, California to the Polar Sea, mostly contained in the collection of the United States National Museum, with illustrations of hitherto unfigured species. Bull. U.S. Nat. Mus. 112: 1-217.

—, 1925. Notes on the nomenclature of our East American species of Pecten with descriptions of new species. Nautilus 38: 112-120.

—, 1925. Illustrations of unfigured types of shells in the collection of the United States National Museum. Proc. U.S. Nat. Mus. 66(17): 1-41.

—, 1926. A new Pecten from the Loochoo Islands. Nautilus 40: 67.

—, 1926. A new Pecten from Colombia. Proc. biol. Soc. Wash. 39: 61.

—, P. Bartsch & H.A. Rehder, 1938. A manual of the Recent and fossil marine Pelecypod mollusks of the Hawaiian Islands. Bernice P. Bishop Mus., Bull. 153: 1-233.

Dance, S.P., 1966. Shell collecting, an illustrated history: 1-344. London.

Dautzenberg, P. & A. Bavay, 1904. Description d'un Amussium nouveau dragué par le Siboga dans la mer de Celebes. J. Conch. Paris 52: 207-211.

— & —, 1912. Les lamellibranches de l'expédition du Siboga. Partie systématique. I. Pectinidés; 1-41. Leiden.

De Castellanos, Z.J., 1967. Catalogo de los moluscos marinos bonaerenses. An. Invest. cient. Prov. B. Aires. 8: 9-365.

—, 1971. Los Chlamys mas comunes del mar Argentino. Neotropica 17: 55-66.

—, 1973. Dos nuevos Chlamys del sur Argentino. Neotropica 19: 62-65.

Defrance, M.J.L., 1825. Dictionnaire des sciences naturelles. 38: 234-267.

Dijkstra, H.H., 1983. Rare or poorly known pectinids. Part 1. Conchiglia (Shell) 176-177:3-5.

—, 1983. Pectinidae de Nouvelle-Calédonie/Pectinidae of New Caledonia. 1. Chlamys (Chlamys) gloriosa (Reeve, 1853). Rossiniana 20: 18-21.

—, 1983. Pectinidae de Nouvelle-Calédonie/Pectinidae of New Caledonia. 2.Scaeochlamys livida peroniana Iredale, 1939. Rossiniana 21: 19-22, illustr.

—, 1984. Rare or poorly known pectinids. Part 2. Conchiglia (Shell) 178-179: 8-9.

—, 1984. Rare or poorly known pectinids. Part 3. Conchiglia (Shell) 182-183: 3-5.

—, 1984. Rare or poorly known pectinids. Part 4. Conchiglia (Shell) 184-185: 4-5.

—, 1984. Rare or poorly known pectinids. Part 5. Conchiglia (Shell) 186-187: 8-9.

—, 1984. Rare or poorly known pectinids. Part 6.
Conchiglia (Shell) 188-189: 28-29.
—, 1984 Pectinidae de Nouvelle-Calédonie/Pectinidae of New Caledonia. 3. Chlamys squamosa (Gmelin, 1791).
Rossiniana 22: 16-17.
—, 1984. Pectinidae de Nouvelle-Calédonie/Pectinidae of New Caledonia. 4. Chlamys coruscans coruscans (Hinds, 1845).
Rossiniana 23: 9-10.
—, 1984. Pectinidae de Nouvelle-Calédonie/Pectinidae of New Caledonia. 5. Comptopallium radula (Linnaeus, 1758).
Rossiniana 24: 11-12.
—, 1984. Pectinidae de Nouvelle-Calédonie/Pectinidae of New Caledonia. 6. Gloripallium pallium (Linnaeus, 1758).
Rossiniana 25: 17-18.
—, 1985. Rare or poorly known pectinids. Part 7.
Conchiglia (Shell) 190-191: 24-25.
—, 1985. Rare or poorly known pectinids. Part 8.
Conchiglia (Shell) 192-193: 20.
—, 1985. Rare or poorly known pectinids. Part 9.
Conchiglia (Shell) 198-199: 24-27.
—, 1985. Pectinidae de Nouvelle-Calédonie/Pectinidae of New Caledonia. 7. Comptopallium vexillum (Reeve, 1853).
Rossiniana 27: 7-9.
—, 1985. Pectinidae de Nouvelle-Calédonie/Pectinidae of New Caledonia. 8. Juxtamusium coudeini (Bavay, 1903).
Rossiniana 28: 10-11.
—, 1985. Pectinidae de Nouvelle-Calédonie/Pectinidae of New Caledonia. 9. Anguipecten lamberti (Souverbie, 1874).
Rossiniana 29: 9, 11.
—, 1986. Pectinidae de Nouvelle-Calédonie/Pectinidae of New Caledonia. 10. Annachlamys kuhnholtzi (Bernardi, 1860).
Rossiana 30: 19, 21.
—, 1986. Pectinidae de Nouvelle-Calédonie/Pectinidae of New Caledonia. 11. Chlamys perfecta (Melvill, 1909).
Rossiniana 31: 9.
—, 1986. Pectinidae de Nouvelle-Calédonie/Pectinidae of New Caledonia. 12. Chlamys (Coralichlamys) madreporarum ([Petit in] Sowerby II, 1842).
Rossiniana 32: 7-8.
Dunker, W., 1882. Index Molluscorum maris Japonici: 1-301. Cassel.
Finlay, H.J., 1927. A further commentary on New Zealand molluscan systematics. Trans. Proc. N.Z. Inst. 57: 320-485.
Fischer, P., 1858. Notes pour servir à la faune malacologique de l'archipel Calédonien. J. Conch. Paris 7: 329-342.
Fleming, C.A., 1951. The genus Pecten in the West Pacific. J. Conch. Paris 90: 276-282.
—, 1955. A new subspecies of scallop from Byron Bay, New South Wales. Aust. zool. 12: 108-109.
—, 1957. The genus Pecten in New Zealand. Palaeont. Bull. Wellington 26: 1-69.
—, 1966. Reprint Marwich's illustrations of New Zealand shells with checklist.
—. 1976. Pecten (Chlamys) suteri Hertlein, 1933, a neglected nom. nov. for Pecten radiatus Hutton (preoccupied). Conch. Sect. Auckl. Inst. & Mus. 1: 28.

Forbes, E., 1844. Report on the Mollusca and Radiata of the Aegean Sea, and their distribution, considered as bearing on geology. Rep. Br. Ass. Advmt sci: 130-193. London.
Gould, A., 1862. Descriptions of new genera and species of shells. Proc. Boston Soc. Nat. Hist. 8: 280-285.
Grau, G., 1952. A new subspecies of Pecten (Plagioctenium) gibbus (Linné). Nautilus 66: 17-19.
—, 1959. Pectinidae of the eastern Pacific. Allan Hancock Pacific Exped. 23: 1-308.
—, 1960. A new Chlamys from the South Pacific. Nautilus 74: 15-18.
Habe, T., 1964. Notes on the species of the genus Amusium (Mollusca). Bull. natn. sci. Mus. Tokyo 7: 1-5.
Hanley, S.C.T., 1842-1856. An illustrated and descriptive catalogue of Recent bivalve shells: 1-392. London.
Hedley, C., 1902. Scientific results of the trawling expedition of H.M. C.S. "Thetis" off the coast of New South Wales in February and March, 1898. Mollusca. Part I. Brachiopoda and Pelecypoda. Mem. Aust. Mus. 4: 287-324.
—, 1909. Mollusca from the Hope Island, North Queensland. Proc. Linn. Soc. N.S.W. 34: 420-466.
Hertlein, L.G., 1931. Changes of nomenclature of some Recent and fossil Pectinidae from Japan, Porto Rico, South America, New Zealand and California. J. Paleont. 5: 367-369.
—, 1935. The Templeton Crocker expedition of the Californian Academy of Sciences, 1932. No. 25, the Recent Pectinidae. Proc. Calif. Acad. Sci. 21: 301-328.
—, 1936. Three new sections and rectifications of some specific names in the Pectinidae. Nautilus 50: 24-27, 54-58.
—, 1940. Additions to the range of Pecten caurinus Gould. Nautilus 54: 68-69.
—, 1969. Family Pectinidae Rafinesque, 1815. In Moore, R.C. (ed.), Treatise on invertebrate paleontology, part N, 1 Mollusca, 6. Bivalvia: N348-N373.
—, 1972. Description of a new species of Chlamys (Mollusca: Pelecypoda) from the Galápagos Islands. Proc. Calif. Acad. Sci. 39: 1-6.
—, & A.M. Strong, 1946. Mollusks from the West coast of Mexico and central America, part III. Zoologica. N.Y. Zool. Soc. 31: 53-76.
—, & —, 1955. Marine mollusks collected during the "Askoy" expedition to Panama, Colombia, and Ecuador in 1941. Bull. Amer. Mus. nat. Hist. 107: 159-318.
Hirase, S., 1936. A collection of Japanese shells with illustrations in natural colours: 1-217. Tokyo.
Iredale, T., 1925. Mollusca from the continental shelf of eastern Australia. Rec. Aust. Mus. 14: 243-270.
—, 1929. Mollusca from the continental shelf of eastern Australia. Rec. Aust. Mus. 17: 157-189.
—, 1929. Queensland molluscan notes, no. 1. Mem. Qd. Mus. 9: 261-297.
—, 1930. Queensland molluscan notes, no. 2. Mem. Qd. Mus. 10: 73-88.
—, 1939. Mollusca, Part I. Gt.. Barrier Reef Exped. 1928-29. Scient Reps. 5: 209-425.
Jensen, A.S., 1912. The Danish Ingolf Expedition. Lamellibranchiata, Part I: 1-119.
Jeffreys, J.G., 1876. New and peculiar Mollusca of the Pecten and Arca families procured in the "Valorous" Expedition. Ann. Mag. nat. Hist. 18.
—, 1879. On the Mollusca procured during the "Lightning" and "Porcupine" Expeditions 1868-1870. Part 2. Proc. zool. Soc. Lond. 1879: 553-588.
Kaas, P. & A.N.C. Ten Broek, 1942. Nederlandse zeemollusken: 1-232. Amsterdam.
Keen, A., 1971. Seashells of tropical West America: 1-1064. Stanford.

Kira, T., 1955. Coloured illustrations of the shells of Japan: 1-205. Osaka.
—, 1962. Shells of the western Pacific in color: 1-224. Tokyo.
Klappenbach, M.A., 1970. Notas malacologicas, 2 Commun. Zool. Mus. Hist. nat. Montev. 132: 1-7.
Kuroda, T. & T. Habe, 1952. Checklist and bibliography of the Recent marine Mollusca of Japan: 1-210. Tokyo.
Kuester H.C. & W. Kobelt, 1842-1859, 1886-1888. Die Gattungen Spondylus und Pecten. In: Martini & Chemnitz: Systematisches Conchylien-Cabinet, vol. 7 (2): 1-128, 129-296. Nürnberg.
Lamarck, J.B., 1819. Histoire naturelle des animaux sans vertèbres: 128-174. Paris.
Lan, T.C., 1980. Rare shells of Taiwan in color: 1-144. Taipei.
Locard, A., 1899. Les coquilles marines au large des côtes de France: 1-198. Paris.
Macpherson, J.H. & C.J. Gabriel, 1961. Marine molluscs of Victoria: 1-475. Melbourne.
Melvill, J.C. & R. Standen, 1906. The Mollusca of the Persian Gulf, Gulf of Oman, and Arabian Sea, with descriptions of new species. Part II. Pelecypoda. Proc. zool. Soc. Lond., 1906: 873-848.
—, 1907. The marine Mollusca of the Scottish national Antarctic Expedition. Trans. R. Soc. 46: 119-157.
—, & E.R. Sykes, 1898. Notes on a second collection of marine shells from the Andaman Islands, with descriptions of new forms of Terebra. Proc. malac. Soc. Lond. 3: 35-48.
Nickles, M., 1950. Mollusques testacés marins de la côte occidentale d'Afrique: 1-256. Paris.
Nordsieck, F., 1969. Die europäischen Meeresmuscheln (Bivalvia): 1-256. Stuttgart.
Oyama, K., 1951. Amusiinae in Japan. Illustrated catalogue of Japanese Shells.
—, 1957. The molluscan shells.
Oliver, P.G., 1982. The Melvill-Tomlin collection, part 11. Pectinacea (Pectinidae). Handlist molluscan coll. Nat. Mus.Wales: 1-17.
Pasteur-Humbert, C., 1962. Les mollusques marins testacés du Maroc. Catalogue non critique. II. Les lamellibranches et les scaphopodes. Trav. Inst. Chérifien: 1-184.
Philippi, E., 1900. Beiträge zur Morphologie und Phylogenie der Lamellibranchien. II. Zur Stammesgeschichte der Pectiniden. Z. dt. geol. Ges. 52: 64-117.
Powell, A.W.B., 1958. Shells of New Zealand: 1-203. Christchurch.
—, 1958. Mollusca of the Kermadec Islands, Part I. Rec. Auckland Inst. Mus. 5: 65-85.
—, 1979. New Zealand Mollusca: 1-500. Auckland.
Reeve, L.A., 1852. Monograph of the genus Pecten. In: Conchologia Iconica 8. London.
Rehder, H.A., 1944. A new Pectinid shell from the Pacific Ocean, with a note on the genus Pallium Schroeter. Nautilus 58: 52-54.
Rios, E.C., 1970. Coastal Brazilian seashells.: 1-255. Rio Grande.
—, 1975. Brazilian marine mollusks iconography: 1-331. Rio Grande.
Risso, A., 1826. Histoire naturelle des principales productions de l'Europe méridionale et particulièrement de celles des environs de Nice et des Alpes Maritimes. 4: 296-299.
Roth, B., 1975. Description of a new species of Pectinid bivalve from the Juan Fernandez Islands, Chile. J. malac. Soc. Aust. 3: 81-87.
—, & E.V. Coan, 1978. Nomenclatural notes on Hinnites giganteus (Gray). Veliger 20: 297-298.
Say, T., 1824. Account of some of the fossil shells of Maryland. J. Acad. nat. Sci. Philad. 1824: 133-136.
Smith, E.A., 1885. Report on the Lamellibranchiata collected by H.M.S. "Challenger", during the years 1873-76. Rep. Sc. Res. Voyage H.M.S. "Challenger", Zoology 13(35): 1-341.

SOOT-RYEN, T., 1960. Pelecypods from Tristan da Cunha. Res. Norw. Sci. Exped. Tristan da Cunha 49: 1-47.

SOUVERBIE, M. & R.P. MONTROUZIER, 1861. Descriptions d'espèces nouvelles de l'archipel Calédonien. J. Conch. Paris 11: 276-287.

—, & —, 1874. Descriptions d'espèces nouvelles de l'archipel Calédonien. J. Conch. Paris 22: 187-201.

SOWERBY (II), G.B., 1847. Thesaurus conchyliorum or monographs of genera of shells. London.

—, 1852. A conchological manuel: 1-337. London.

SOWERBY (III), G.B., 1904. Mollusca of South Africa (Pelecypoda). Marine Investigations in South Africa 4: 1-19.

—, 1908. Descriptions of eight new species of marine Mollusca. Proc. malac. Soc. Lond. 8: 16-19.

STEP, E., 1951. Shell life, introduction to British Mollusca: London.

TATE, R., 1887. On the Australian Pectens confounded with the New Zealand Pecten laticostatus. Pap. Proc. R. Soc. Tasm. for 1886: 113-116.

THIELE, J., 1934. Handbuch der systematischen Weichtierkunde. 3e Teil. Scaphopoda, Bivalvia, Cephalopoda: 779-1022. Stuttgart.

TRYON, G.W., 1870. Descriptions of new species of marine bivalve Mollusca in the collection of the Academy of Natural Sciences. Am. J. Conch. 6: 23-24.

TURTON, W.H., 1932. The marine shells of Port Alfred, S. Africa: 1-331. Oxford.

VAN BRUGGEN, A.S., 1961. Notes on South African marine Mollusca. 4. A new variety of Pecten sulcicostatus. Basteria 25: 31-32.

VERRILL, A. E., 1873. Results of recent dredging expeditions on the coast of New England. Am. J. Sci. 3(5): 131-138.

—, 1882. Catalogue of marine Mollusca added to the fauna of New England during the past ten years. Trans. Conn. Acad. Arts Sci. 5: 447-587.

—, 1897. A study of the family Pectinidae, with a revision of the genera and sub-genera. Trans. Conn. Acad. Arts Sci. 10: 41-96.

VERRILL, A.E. & K.J. BUSH, 1898. Revision of the deep-water Mollusca of the Atlantic coast of North America, with descriptions of new genera and species. Part I. Proc. U.S. nat. Mus. 20: 775-901.

VOKES, H.E., 1980. Genera of the Bivalvia. A systematic and bibliographic catalogue: 1-307. New York.

WAGNER, H.P., 1982. Notes on type material of the family Pectinidae (Mollusca: Bivalvia) 1. Pecten limatula Reeve, 1853, a new synonym of Chlamys irregularis (Sowerby, 1842). Basteria 46: 86.

—, 1983. Notes on type material of the family Pectinidae (Mollusca: Bivalvia) 2. Notes on Chlamys humilis Sowerby, 1904 and Chlamys natalensis Sowerby, 1906. Basteria 47: 145-148.

WALLER, T.R., 1969. The evolution of the Argopecten gibbus stock (Mollusca: Bivalvia) with emphasis on the Tertiary and Quarternary species of eastern North America. J. Paleont. Soc. Mem. 3(43): 1-125.

—, 1972. The Pectinidae (Mollusca: Bivalvia) of Eniwetok Atoll, Marshall Islands. Veliger 14: 221-264.

WARMKE, G.L. & R.T. ABBOTT, 1961. Caribbean Seashells: 1-346. Narberth.

WEBB, J.H., 1956. A new species of Pecten (Chlamys) from Ryukyu. Venus 19: 22-24.

—, 1957. New form of Pecten. Nautilus 71: 53-54.

WOODWARD, S.P., 1875. A manual of the Mollusca: 1-542. London.

ZOLOTAREV, V.N., 1984. Investigations on the individual growth of marine bivalves. Malacol. Review 17: 126-127.

APPENDIX

LIST OF NEWLY DESCRIBED GENERA, SUBGENERA, SPECIES AND NEW NAMES, NOT TREATED IN ROMBOUTS' MANUSCRIPT.

PECTINIDAE

Bathypecten Schein-Fatton, 1985.
Type species: *Bathypecten vulcani* Schein-Fatton, 1985.

Crassodoma Bernard, 1986.
Type species: *Lima gigantea* Gray, 1825 = *Hinnites giganteus* (Gray, 1925).

Somalipecten Waller, 1986.
Type species: *Somalipecten cranmerorum* Waller, 1986.

Karnekampia H.P. Wagner, 1988.
Type species: *Pecten bruei* Payraudeau, 1826.

Perapecten H.P. Wagner, 1985. Subgenus of *Aequipecten*.
Type species: *Pecten commutatus* Monterosato, 1875.

Dentamussium Dijkstra, 1990. Subgenus of *Amusium*.
Type species: *Ostrea obliteratum* Linnaeus, 1758.

Delectopecten musorstomi Poutiers, 1981.
Type locality: 13 - 57.0 N (North of Lubang, Philippines).
120 - 16.5 E

Pecten berry Bernard, 1983.
nomen novum for *Pecten lunaris* Berry, 1963 (not Roemer, 1839).

Cryptopecten complanus Zhenrui, 1983.
Type locality: 31 - 05 N (East China Sea).
128 - 00 E
(= *Pecten (Chlamys) bullatus* Dautzenberg & Bavay, 1912).

Bractechlamys elegans Zhenrui, 1983.
Type locality: China, Yülin, Hainan Island.
(= *Pecten quadriliratus* Lischke, 1870).

Hinnites spectabilis Cosel & Gofas, 1984.
Type locality: Angola, Baia de Santa Maria, Benguela.
(= *Hinnites corallinus* Sowerby I, 1827)

Bathypecten vulcani Schein-Fatton, 1985.
Type locality: 12 - 48.8 N (East Pacific)
103 - 56.6 W

Somalipecten cranmerorum Waller, 1986.
Type locality: East Africa, off Somalia.

Semipecten kengaluorum Dijkstra, 1986.
Type locality: Solomon Islands, Russell Island.

Chlamys allorenti Dijkstra, 1988.
Type locality: 21 - 01 S (Reunion, off Boucan Canot).
55 - 13 E

Mirapecten moluccensis Dijkstra, 1988.
Type locality: Indonesia, Moluccas.

Chlamys elsae H.P. Wagner, 1988.
Type locality: Philippines, Bohol, Panglao.

Chlamys aliae Dijkstra, 1988.
Type locality: off Punta Engano, Philippines.

Semipallium barnetti Dijkstra, 1988.
Type locality: off Punta Engano, Philippines.

Pseudohinnites levii Dijkstra, 1989.
Type locality: S.E. of New Caledonia.

Palliolum tydemani Dijkstra, 1990.
Type locality: Ceram Sea, Indonesia.

PROPEAMUSSIIDAE

Cyclopecten argenteus Bernard, 1978.
Type locality: 44 - 25.6 N (USA Oregon, West of Cape Blanco.).
125 - 13.4 W

Cyclopecten knudseni Bernard, 1978.
Type locality: 45 - 17.4 N (USA., Oregon, West of Yaquina Head.).
125 - 49.0 W

Cyclopecten squamiformis Bernard, 1978.
Type locality: 45 - 52.7 N (USA, Oregon, West of Astoria.).
126 - 29.4 W

Ctenamussium sinensis Zhenrui, 1980.
Type locality: 26 - 30 N (East China Sea).
124 - 00 E

Propeamussium (Propeamussium) stella Zhenrui, 1984.
Type locality: 19.0 - 00 N (South China Sea).
112.5 - 00 E

Propeamussium (Propeamussium) gracilis Zhenrui, 1984.
Type locality: 19 - 00 N (South China Sea)
113 - 00 E

Propeamussium dautzenbergi Dijkstra, 1990.
Type locality: Macassar Strait, Indonesia.

Cyclopecten bavayi Dijkstra, 1990.
Type locality: Ceram Sea, Indonesia.

Cyclopecten brundisiensis Smriglio & Mariottini, 1990.
Type locality: Brindisi, Italy.

PLATES

The abbreviations, used to indicate collections, are:

AH = A. Herlaar
AMS = Australian Museum Sydney
AR = A. Rombouts (now in ZMA)
HD = Henk H. Dijkstra
JB = Jan A. Buijse
LACM = Los Angeles County Museum
ZMA = Zoological Museum Amsterdam
PP = Peter L. van Pel

Plate 1

1. *Aequipecten audouinii* (Paryaudeau, 1826)
   30 mm; Adriatic Sea
   (coll. ZMA ex AR) See p. 1

2. *Aequipecten opercularis* (Linnaeus, 1758)
   41 mm; France, Bretagne
   (coll. PP) See p. 2

2a. idem, 50 mm

2b. idem, 42 mm

3. *Aequipecten glyptus* (Verrill, 1882)
   57 mm; USA., Florida, Dry Tortugas
   (coll. ZMA ex AR) See p. 1

4. *Aequipecten phrygius* (Dall, 1886)
   37 mm; USA., Florida Straigts.
   (coll. HD) See p. 3

5. *Aequipecten muscosus* (Wood, 1828)
   41 mm; USA, Florida
   (coll. ZMA ex AR) See p. 2

5a. 34 mm; USA, Florida, Key West
   (coll. HD)

6. *Aequipecten palmeri* (Dall, 1897)
   33 mm; Mexico
   (coll. ZMA ex AR) See p. 3

7. *Aequipecten lineolaris* (Lamarck, 1819)
   32 mm; Honduras
   (coll. ZMA ex AR) See p. 2

1
2
2 a
2 b
3
4
5
5 a
6
7

Plate 2

1. *Adamussium colbecki* (Smith, 1902)
   82 mm; Antarctica Ross Sea
   (coll. HD) See p. 1

2. *Amusium japonicum* (Gmelin, 1791)
   100 mm; Japan
   (coll. ZMA ex AR) See p. 4

3. *Amusium obliteratum* (Linnaeus, 1758)
   58 mm; Indonesia, Moluccas
   (coll. ZMA) See p. 4

4. *Amusium laurentii* (Gmelin, 1791)
   55 mm; Honduras
   (coll. ZMA ex AR) See p. 4

5. *Amusium balloti* (Bernardi, 1861)
   95 mm; Australia, W.A.
   (coll. ZMA ex AR) See p. 3

5a. 100 mm; Australia, N.S.W.
   (coll. ZMA ex AR)

1
2
3
4
3
5
5 a

Plate 3

1. *Amusium papyraceum* (Gabb, 1873)
   52 mm; USA Florida, Tampa Bay
   (coll. ZMA ex AR) See p. 4

2. *Argopecten commutatus* (Monterosato, 1875)
   27 mm; North Africa, Goree
   (coll. ZMA ex AR) See p. 6

3. *Amusium pleuronectes* (Linnaeus, 1758)
   80 mm; Indonesia, Moluccas
   (coll. ZMA) See p. 4

4. *Argopecten circularis* (Sowerby, 1835)
   61 mm; Mexico, Cholla Bay
   (coll. ZMA ex AR) See p. 5

4a. 53 mm; Panama
   (coll. ZMA ex AR)

5. *Argopecten flabellum* (Gmelin, 1791)
   45 mm; North West Africa
   (coll. PP) See p. 6

5a. 33 mm; West Africa, Sierra Leone
   (coll. ZMA ex AR)

6. *Argopecten gibbus* (Linnaeus, 1758)
   35 mm; USA Florida, Panama Key
   (coll. PP) See p. 6

6a. 36 mm; USA Florida, Sanibel Isl.
   (coll. PP)

1
2
3
4
4 a
5
5 a
6
6 a

Plate 4

1. *Argopecten irradians* (Lamarck, 1819)
   51 mm; USA., Georgia
   (coll. ZMA ex AR) See p. 6

1a. 50 mm; USA, Florida
   (coll. ZMA ex AR)

2. *Argopecten irradians concentricus* (Say, 1822)
   51 mm; USA., Florida, Panama City
   (coll. ZMA ex AR) See p. 7

3. *Argopecten nucleus* (Born, 1778)
   30 mm; USA, Florida
   (coll. ZMA) See p. 7

4. *Chlamys cloacata* (Reeve, 1853)
   33 mm; Japan, Mikawa
   (coll. ZMA ex AR) See p. 10

5. *Argopecten purpuratus* (Lamarck, 1819)
   76 mm; Peru, Callao
   (coll. ZMA) See p. 8

6. *Chlamys islandica erythrocomata* (Dall, 1907)
   79 mm; U.S.S.R., Vladivostok
   (coll. ZMA ex AR) See p. 14

1
1 a
2
3
4
5
6

Plate 5

1. *Chlamys cristularis* (Adams & Reeve, 1850)
   56 mm; China
   (coll. ZMA ex AR) See p. 10

2. *Chlamys dichroa* (Suter, 1909)
   42 mm; New Zealand, South Island
   (coll. HD) See p. 11

3. *Semipallium dianae* (Crandall, 1979)
   43 mm; Japan
   (coll. HD) See p. 59

3a. 38 mm; Japan, Okinawa
   (coll. HD)

3b. idem, 46 mm

4. *Chlamys dieffenbachi* (Reeve, 1853)
   49 mm; New Zealand, Steward Island
   (coll. ZMA ex AR) See p. 11

5. *Chlamys hindsii* (Carpenter, 1864)
   51 mm; USA, Puget Sound
   (coll. ZMA) See p. 13

6. *Volachlamys fultoni* (Sowerby, 1904)
   30 mm; Mozambique
   (coll. HD) See p. 62

7. *Chlamys farreri* (Jones & Preston, 1904)
   64 mm; Japan
   (coll. ZMA ex AR) See p. 12

8. *Chlamys hastata* (Sowerby, 1842)
   45 mm; USA, California, San Diego
   (coll. ZMA ex AR) See p. 12

8a. 65 mm; US.A., Oregon
   (coll. ZMA ex AR)

1
2
3
3 a
3 b
4
5
6
7
8
8 a

Plate 6

1. *Chlamys lischkei* (Dunker, 1850)
   58 mm; Uruguay
   (coll. PP) See p. 35

2. *Chlamys lemniscata* (Reeve, 1853)
   33 mm; Japan, Tosa Bay
   (coll. ZMA ex AR) See p. 15

3. *Argopecten pelseneeri* (Dautzenberg & Bavay, 1912)
   26 mm; Japan, Mikawa
   (coll. ZMA ex AR) See p. 7

4. *Chlamys albolineata* (Sowerby, 1842)
   25 mm; Japan, Okinawa
   (coll. ZMA ex AR) See p. 9

5. *Chlamys jousseaumei* Bavay, 1904
   26 mm; Japan, Kii
   (coll. ZMA ex AR) See p. 14

6. *Chlamys mildredae* (Bayer, 1941)
   35 mm; USA Florida, Pompano Beach
   (coll. HD) See p. 16

7. *Chlamys kiwaensis* Powell, 1933
   27 mm; New Zealand, Cape Campbell
   (coll. HD) See p. 15

8. *Chlamys mollita* (Reeve, 1853)
   36 mm; Papua New Guinea, Milne Bay
   (coll. HD) See p. 16

9. *Chlamys incantata* Hertlein, 1972
   47 mm; Galapagos Islands
   (coll. HD) See p. 13

10. *Chlamys multisquamata* (Dunker, 1864)
    50 mm; Netherlands Antilles, Curaçao
    (coll. HD) See p. 16

10a. idem, 46 mm
     (coll. PP)

1
2
3
4
5
6
7
8
9
10
10a

Plate 7

1. *Chlamys nivea* (MacGillivray, 1825)
   52 mm; U.K., Scotland
   (coll. PP) See p. 16

2. *Chlamys ornata* (Lamarck, 1819)
   27 mm; Netherlands Antilles, Curaçao
   (coll. PP) See p. 17

2a. idem, 39 mm

3. *Chlamys phalara* Roth, 1975
   26 mm; Chili, Juan Fernandez Island
   (coll. HD) See p. 17

4. *Chlamys rubida* (Hinds, 1845)
   49 mm; USA Puget Sound
   (coll. ZMA ex AR) See p. 18

5. *Chlamys ruschenbergerii* (Tryon, 1869)
   52 mm; Oman, Muscat
   (coll. ZMA ex AR) See p. 18

5a. idem, 55 mm

6. *Chlamys squamata* (Gmelin, 1791)
   50 mm; Japan, Mikawa
   (coll. ZMA ex AR) See p. 19

6a. idem, 54 mm

7. *Chlamys taiaroa* Powell, 1952
   52 mm; New Zealand, Dunedin
   (coll. ZMA ex AR) See p. 19

8. *Chlamys rubromaculata* (Sowerby, 1842)
   34 mm; Gulf of Aqaba
   (coll. ZMA ex AR) See p. 18

9. *Chlamys tehuelcha* (d'Orbigny, 1846)
   49 mm; Uruguay, Nastrei
   (coll. ZMA ex AR) See p. 19

1
2
2 a
3
4
5
5 a
6
6 a
7
8
9

Plate 8

1. *Chlamys varia* (Linnaeus, 1758)
   36 mm; France Bretagne
   (coll. PP) See p. 20

1a. idem, 31 mm

1b. idem, 35 mm

2. *Chlamys zeelandona* (Hertlein, 1931)
   29 mm; New Zealand
   (coll. HD) See p. 21

2a. 23 mm; New Zealand, Goat Island.
   (coll. ZMA ex AR)

3. *Chlamys zeelandiae* (Gray, 1843)
   28 mm; New Zealand, Nelson
   (coll. ZMA ex AR) See p. 21

4. *Chlamys oweni* (De Gregorio, 1884)
   22 mm; Japan, Okinawa
   (coll. HD) See p. 24

5. *Chlamys aktinos* (Petterd, 1886)
   34 mm; Australia, W.A., Albany
   (coll. PP) See p. 22

6. *Chlamys vesiculosa* (Dunker, 1877)
   27 mm; Japan, Kii
   (coll. ZMA ex AR) See p. 24

6a. idem, 29 mm

1
1 a
1 b
2
2 a
3
4
5
6
6 a

Plate 9

1. *Chlamys townsendi* (Sowerby, 1895)
   116 mm; Gulf of Oman
   (coll. ZMA ex AR) See p. 20

2. *Chlamys corallina* (Sowerby, 1827)
   67 mm; Angola, Mossamedes
   (coll. HD) See p. 24

3. *Chlamys ercoliniana* (Cocconi, 1873)
   43 mm; Senegal, Cape Manuel
   (coll. HD) See p. 24

4. *Chlamys gigantea* (Gray, 1825)
   59 mm; USA, California, Long Beach
   (Coll. PP) See p. 25

5. *Chlamys pusio* (Linnaeus, 1758)
   37 mm; U.K., Wales
   (coll. PP) See p. 25

6. *Chlamys leopardus* (Reeve, 1853)
   56 mm; Australia, QLD., Tin Can Bay
   (coll. PP) See p. 22

7. *Chlamys macassarensis* (Chenu, 1845)
   42 mm; Philippines, Zamboanga
   (coll. ZMA ex AR) See p. 22

8. *Chlamys macassarensis* fa *reevei* (Adams & Reeve, 1850)
   47 mm; Philippines, Sulu Sea
   (coll. HD) See p. 22

1
2
3
4
5
6
7
8

Plate 10

1. *Chlamys blanda* (Reeve, 1853)
   29 mm; Australia, QLD., Turkey Beach
   (coll. PP) See p. 27

2. *Chlamys latiaurata* (Conrad, 1837)
   26 mm; USA., California, San Diego
   (coll. ZMA ex AR) See p. 26

2a. idem, 32 mm

3. *Chlamys asperrima* (Lamarck, 1819)
   66 mm; Australia, S.A.
   (coll. ZMA ex AR) See p. 27

3a. idem, 78 mm

3b. idem, 71 mm

4. *Chlamys australis* (Sowerby, 1842)
   78 mm; Australia, W.A., Shark Bay
   (coll. PP) See p. 27

1
2
2 a
3
3 a
3 b
4

Plate 11

1. *Chlamys lentiginosa* (Reeve, 1853)
   52 mm; Papua New Guinea, Badu Island
   (coll. ZMA ex AR) See p. 29

1a. 49 mm; Indonesia, Moluccas
   (coll. ZMA ex AR)

2. *Chlamys funebris* (Reeve, 1853)
   43 mm; Australia, W.A., Northwest coast
   (coll. ZMA ex AR) See p. 28

2a. 55 mm; Australia, W.A. Cape Leveque
   (coll. PP)

3. *Chlamys gemmulata* (Reeve, 1853)
   45 mm; New Zealand, off Dunedin
   (coll. ZMA ex AR) See p. 28

4. *Chlamys gloriosa* (Reeve, 1853)
   84 mm; Australia, QLD.
   (coll. ZMA ex AR) See p. 29

4a. idem, 58 mm

4b. 37 mm; Australia, N.T., Gulf of Carpenteria
   (coll. ZMA ex AR)

4c. 52 mm; Papua New Guinea, Badu Island
   (coll. ZMA ex AR)

5. *Chlamys scabricostata* (Sowerby, 1915)
   53 mm; Australia, W.A., Broome
   (coll. ZMA ex AR) See p. 29

6. *Chlamys senatoria* (Gmelin, 1791)
   38 mm; Philippines, Zamboanga
   (coll. ZMA ex AR) See p. 30

6a. 38 mm; Philippines, Samar
   (coll. PP)

1
2
2 a
3
4
4 a
1 a
4 b
5
6
6 a
4c

Plate 12

1. *Chlamys sentis* (Reeve, 1853)
35 mm; Mexico
(coll. ZMA ex AR) See p. 30

2. *Chlamys squamosa* (Gmelin, 1791)
42 mm; Australia, QLD.
(coll. ZMA ex AR) See p. 30

3. *Chlamys mirifica* Reeve (1853)
32 mm; Philippines, Bohol
(coll. HD) See p. 31

3a. 40 mm; Philippines, Cebu
(coll. HD)

4. *Chlamys rastellum* (Lamarck, 1819)
34 mm; Philippines, Cebu
(coll. HD) See p. 31

4a. idem, 34 mm

4b. idem, 35 mm

5. *Chlamys thaanumi* (Dall, Bartsch & Rehder, 1938)
35 mm; Red Sea
(coll. HD) See p. 31

6. *Chlamys linki* (Dall, 1926)
27 mm; Gulf of Mexico
(coll. HD) See p. 32

7. *Chlamys tumbezensis* (D'Orbigny, 1846)
31 mm; Mexico, Mazatlan
(coll. ZMA ex AR) See p. 32

8. *Chlamys irregularis* (Sowerby, 1842)
27 mm; Japan
(coll. ZMA ex AR) See p. 33

9. *Chlamys livida* (Lamarck, 1819)
49 mm; Australia, NSW
(coll. ZMA ex AR) See p. 33

1
2
3
4
4 a
4 b
5
3 a
6
7
8
9

Plate 13

1. *Chlamys swiftii* (Bernardi, 1858)
   79 mm; Japan
   (coll. ZMA ex AR) See p. 33

2. *Chlamys delicatula* (Hutton, 1873)
   73 mm; New Zealand, off Dunedin
   (coll. ZMA ex AR) See p. 34

3. *Chlamys patagonica* (King & Broderip, 1832)
   45 mm; Argentina Patagonia, Orange Harbour
   (coll. ZMA) See p. 35

4. *Comptopallium radula* (Linnaeus, 1758)
   75 mm; Australia, W.A., Pardoo Sands
   (coll. ZMA ex AR) See p. 35

5. *Comptopallium radula griggi* (Webb, 1957)
   78 mm; (right valve figured) Australia, W.A., North of Broome
   (coll. HD) See p. 35

6. *Corymbichlamys iredalei* (Powell, 1958)
   75 mm; New Caledonia, off Noumea
   (coll. HD) See p. 36

7. *Decadopecten amiculum* (Philippi, 1851)
   40 mm; Mozambique
   (coll. ZMA ex AR) See p. 36

8. *Decadopecten plica* (Linnaeus, 1758)
   39 mm; Japan, Mikawa
   (coll. ZMA ex AR) See p. 37

8a. idem, 35 mm

9. *Decadopecten strangei* (Reeve, 1852)
   46 mm; Australia, QLD., Moreton Bay
   (coll. ZMA ex AR) See p. 37

9a idem, 44 mm

10. *Decadopecten subplicatus* (Sowerby, 1842)
    53 mm; Indonesia
    (coll. ZMA) See p. 37

11. *Decadopecten velutinus* (Sowerby, 1842)
    36 mm; Philippines
    (coll. ZMA ex AR) See p. 37

12. *Decadopecten aurantiacus* (Adams & Reeve, 1850)
    42 mm; Japan, Okinawa, Seragaki area
    (coll. HD) See p. 38

1
2
3
4
5
6
7
8
8a
9
9a
10
11
12

Plate 14

1. *Decadopecten lamberti* (Souverbie & Montrouzier, 1874)
   65 mm; New Caledonia, Boulardi Pass
   (coll. HD) See p. 38

2. *Decadopecten superbus* (Sowerby, 1842)
   60 mm; Japan, Okinawa, Seragaki
   (coll. HD) See p. 38

2a. 47 mm; Indonesia Amboina
   (coll. ZMA)

3. *Equichlamys bifrons* (Lamarck, 1819)
   81 mm; Australia, Tasmania
   (coll. ZMA ex AR) See p. 38

3a. idem, 85 mm

4. *Euvola raveneli* (Dall, 1898)
   54 mm; USA, Florida
   (coll. HD) See p. 39

4a. idem, 41 mm
   (coll. ZMA ex AR)

4b. idem, 41 mm
   (coll. HD)

5. *Euvola ziczac* (Linnaeus, 1758)
   51 mm; Gulf of Venezuela, SW of Islas Las Monjes
   (coll. HD) See p. 39

5a. 43 mm; USA Florida, Key West

5b. 56 mm; Bermuda, Truck Isl.
   (coll. ZMA ex AR)

6. *Excellichlamys histrionica* (Gmelin, 1791)
   32 mm; Thailand
   (coll. HD) See p. 39

7. *Excellichlamys spectabilis* (Reeve, 1853)
   32 mm; New Caledonia, Noumea
   (coll. HD) See p. 40

8. *Flexopecten pesanatis* (Reeve, 1852)
   50 mm; Indonesia, Moluccas
   (coll. ZMA) See p. 42

1
2
2 a
3
3 a
4
4 a
4 b
5
5 a
5 b
6
7
8

Plate 15

1. *Flexopecten glaber* (Linnaeus, 1758)
   38 mm; Adriatic Sea, Dalmatia
   (coll. ZMA ex AR) See p. 41

1a. 43 mm; Turkey, Sea of Marmora
   (coll. ZMA ex AR)

1b. 50 mm; Greece, Mytilini
   (coll. ZMA ex AR)

2. *Flexopecten unicolor* (Lamarck, 1819)
   43 mm; Adriatic Sea
   (coll. ZMA ex AR) See p. 42

2a. idem, 60 mm

3. *Flexopecten proteus* (Solander in Dillwyn, 1817)
   51 mm; Italy
   (coll. HD) See p. 42

4. *Gloripallium pallium* (Linnaeus, 1758)
   41 mm; Philippines
   (coll. PP) See p. 42

4a. idem, 55 mm

5. *Gloripallium sanguinolentum* (Gmelin, 1791)
   43 mm; Gulf of Aquaba
   (coll. HD) See p. 43

5a. 45 mm; Red Sea
   (coll. ZMA ex AR)

6. *Gloripallium spiniferum* (Sowerby, 1835)
   28 mm; Tuamotu Islands.
   (coll. HD) See p. 43

6a. idem, 25 mm

7. *Haumea juddi* Dall, Bartsch & Rehder, 1938
   31 mm; USA, Hawaii Islands
   (coll. HD) See p. 44

8. *Gloripallium speciosum* (Reeve, 1853)
   52 mm; Japan, Okinawa, Seragaki
   (coll. HD) See p. 43

8a. idem, 40 mm

8b. idem, 56 mm

8c. idem, 34 mm

1
1 a
1 b
2
2 a
3
4
4 a
5
5 a
6
6 a
7
8
8 a
8 b
8 c

Plate 16

1. *Manupecten pesfelis* (Linnaeus, 1758)
   59 mm; Senegal, Cape Manuel
   (coll. HD) See p. 46

1a. 39 mm; Italy
   (coll. HD)

2. *Lissopecten hyalinus* (Poli, 1795)
   23 mm; Italy, Elba
   (coll. HD) See p. 44

3. *Manupecten alicei* (Dautzenberg & Fischer, 1897)
   21 mm; South Africa, Namibia
   (coll. HD) See p. 46

4. *Lyropecten corallinoides* (D'Orbigny, 1834)
   37 mm; Canary Islands, La Palma
   (coll. HD) See p. 45

5. *Nodipecten langfordi* Dall, Bartsch & Rehder, 1938
   32 mm; USA, Hawaii, Oahu
   (coll. HD) See p. 46

6. *Nodipecten noduliferus* (Sowerby, 1842)
   47 mm; off Somalia
   (coll. HD) See p. 47

6a. idem, 47 mm

7. *Notochlamys anguinea* (Finlay, 1927)
   39 mm; Australia, W.A. Albany
   (coll. HD) See p. 47

7a. 39 mm; Australia, W.A., Esperance
   (coll. HD)

7b. 34 mm; Australia, W.A., Albany
   (coll. HD)

7c. 42 mm (innerside); Australia, W.A., Emu Point, Albany
   (coll. PP)

1
1 a
2
3
4
5
6
6 a
7
7 a
7 b
7 c

Plate 17

1. *Lyropecten subnodosus* (Sowerby, 1835)
110 mm; Mexico
(coll. HD) See p. 45

1a. idem, 95 mm
(coll. ZMA ex AR)

2. *Nodipecten nodosus* (Linnaeus, 1758)
104 mm; Gulf of Mexico
(coll. ZMA ex AR) See p. 47

2a. 76 mm; USA., Florida, Dry Tortugas
(coll. ZMA ex AR)

1
1 a
2
2 a

Plate 18

1. *Notochlamys tasmanica* (Adams & Angas, 1863)
   37 mm; South Australia
   (coll. HD) See p. 48

2. *Palliolum tigerinum* (Müller, 1776)
   28 mm; S.E. Ireland, Nymphe Bank
   (coll. ZMA) See p. 49

2a. idem, 27 mm

2b. idem, 25 mm

3. *Pecten albicans* Schröter, 1802
   38 mm; Japan, Kagoshima
   (coll. ZMA ex AR) See p. 49

3a. idem, 39 mm

4. *Pecten chazaliei* Dautzenberg, 1900
   27 mm; Mexico, Yucatan
   (coll. HD) See p. 49

4a. 30 mm; Br. Guyana
   (coll. HD)

5. *Pecten diegensis* Dall, 1898
   76 mm; USA California, Redondo Beach
   (coll. ZMA ex AR) See p. 49

6. *Pecten dorotheae* Melvill & Standen, 1907
   60 mm; Somalia
   (coll. PP) See p. 50

7. *Pecten excavatus* Anton, 1839
   50 mm; Japan
   (coll. ZMA) See p. 50

8. *Pecten fumatus* Reeve, 1852
   59 mm; Australia, QLD., Fraser Island
   (coll. ZMA ex AR) See p. 50

1
2
2a
2b
3
3a
4
4a
5
6
7
8

Plate 19

1. *Pecten jacobaeus* (Linnaeus, 1758)
   51 mm; Italy, Naples
   (coll. PP) See p. 51

1a. 69 mm; Italy, off Grado
   (coll. HD)

2. *Pecten maximus* (Linnaeus, 1758)
   83 mm; North Sea
   (coll. ZMA) See p. 51

3. *Pecten meridionalis* Tate, 1887
   87 mm; Australia, Tasmania
   (coll. ZMA ex AR) See p. 52

4. *Pecten medius* Lamarck, 1819
   53 mm; New Zealand
   (coll ZMA ex AR) See p. 51

5. *Pecten novaezelandicus* Reeve, 1852
   68 mm; New Zealand, Averland
   (coll. ZMA ex AR) See p. 52

6. *Pecten sericeus* Hinds, 1845
   51 mm; Mexico, Sonora
   (coll. ZMA ex AR) See p. 52

7. *Pecten sinensis* Sowerby, 1842
   47 mm; Japan, Tosa Bay
   (coll. ZMA ex AR) See p. 52

8. *Pecten sulcicostatus* Sowerby, 1842
   96 mm; South Africa, False Bay
   (coll. HD) See p. 53

1
1 a
2
3
4
5
6
7
8

Plate 20

1. *Pecten waikikius* Dall, Bartsch & Rehder, 1938
   28 mm; USA Hawaii, Oahu
   (coll. HD) See p. 53
2. *Pecten pyxidatus* (Born, 1778)
   45 mm; Philippines, Cebu
   (coll. PP) See p. 53
3. *Pecten erythraeensis* Sowerby, 1842
   47 mm; Red Sea
   (coll. HD) See p. 54
4. *Pecten perulus* Olsson, 1961
   26 mm; Panama, Cebaeo Isl.
   (coll. ZMA ex AR) See p. 54
5. *Pecten vogdesi* Arnold, 1906
   62 mm; USA California, Los Angeles
   (coll. ZMA ex AR) See p. 54
6. *Pecten gardineri* (Smith, 1903)
   30 mm; Taiwan
   (coll. HD) See p. 55
7. *Pecten tricarinatus* Anton, 1839
   30 mm; Philippines, Quezon
   (coll. ZMA ex AR) See p. 55

7a. idem, 30 mm

8. *Pseudamussium hybridum* (Gmelin, 1791)
   31 mm; Senegal
   (coll. HD) See p. 57
9. *Pseudamussium septemradiatum* (Müller, 1776)
   46 mm; Norway, Tromsô
   (coll. ZMA ex AR) See p. 57
10. *Pseudamussium clavatum* (Poli, 1795)
    27 mm; Mediterranean Sea
    (coll. ZMA) See p. 57
11. *Pseudamussium fasciculatum* (Hinds, 1845)
    32 mm; Mexico
    (coll. HD) See p. 58
12. *Semipallium crouchi* (Smith, 1892)
    49 mm; Gulf of Eilat
    (coll. HD) See p. 58
13. *Semipallium fulvicostatum* (Adams & Reeve, 1850)
    28 mm; Tahiti, Arue
    (coll. HD) See p. 59
14. *Semipallium imbricatum* (Gmelin, 1791)
    26 mm (innerside); Netherlands Antilles, Curaçao
    (coll. PP) See p. 59

14a. idem, 32 mm

1
2
3
4
5
6
7
7a
8
9
10
11
12
13
14
14a

Plate 21

1. *Patinopecten caurinus yessoensis* (Jay, 1839)
   139 mm; Japan
   (coll. ZMA) See p. 56

2. *Placopecten magellanicus* (Gmelin, 1791)
   115 mm; North Atlantic Ocean
   (coll. HD) See p. 56

3. *Patinopecten caurinus* (Gould, 1850)
   97 mm; USA Washington, off Cape Flattery
   (coll. ZMA ex AR) See p. 55

3a. 134 mm; USA Washington
   (coll. ZMA ex AR)

1
2
3
3 a

Plate 22

1. *Semipallium tigris* (Lamarck, 1819)
   42 mm; Guam
   (coll. HD) See p. 59

1a. 52 mm (innerside); New Guinea, Biak
   (coll. PP)

2. *Semipallium vexillum* (Reeve, 1853)
   31 mm; Philippines, Cebu
   (coll. PP) See p. 60

2a. idem, 40 mm

3. *Mesopeplum convexum* (Quoy & Gaimard, 1835)
   50 mm; New Zealand
   (coll. HD) See p. 61

4. *Mesopeplum caroli* Iredale, 1929
   42 mm; Australia, QLD.
   (coll. HD) See p. 61

5. *Volachlamys hirasei* (Bavay, 1904)
   47 mm; Japan, Kagawa
   (coll. HD) See p. 62

6. *Volachlamys singaporina* (Sowerby, 1842)
   37 mm; South China Sea, East of Malacca
   (coll. HD) See p. 62

7. *Volachlamys cumingii* (Reeve, 1853)
   56 mm; Australia, QLD., Port Alma
   (coll. ZMA ex AR) See p. 61

7a. 38 mm; Australia, QLD., Moreton Bay
   (coll. ZMA ex AR)

8. *Volachlamys tranquebaria* (Gmelin, 1791)
   38 mm; India, Madras
   (coll. PP) See p. 63

8a. idem, 39 mm

1
1 a
2
2 a
3
4
5
6
7
7 a
8
8 a

Plate 23

1. *Aequipecten minutus* (Nicklès, 1955)
   14 mm; Angola, Luanda
   (coll. HD) See p. 2

2. *Argopecten asperulatus* (Adams & Reeve, 1850)
   19 mm; Japan, Wakayama
   (coll. HD) See p. 5

3. *Chlamys bernardi* (Philippi, 1851)
   13 mm; Philippines, Cebu
   (coll. PP) See p. 23

3a. idem, 15 mm

4. *Chlamys benedicti* Verrill & Busch, 1897
   13 mm; USA, Florida, Cape San Blas
   (coll. ZMA ex AR) See p. 9

5. *Chlamys bruei* (Payraudeau, 1826)
   16 mm; Italy
   (coll. HD) See p. 9

6. *Chlamys bullata* (Dautzenberg & Bavay, 1912)
   19 mm; Japan, Tosa Bay
   (coll. ZMA ex AR) See p. 23

7. *Chlamys coruscans* (Hinds, 1845)
   14 mm; Australia, QLD., Tryon Island
   (coll. HD) See p. 10

7a. idem, 18 mm

7b. idem, 16 mm

8. *Chlamys empressae* Kuroda & Habe, 1971
   17 mm; Philippines, Cebu
   (coll. HD) See p. 11

8a. idem, 15 mm

1
2
3
3 a
4
5
6
7
7 a
7 b
8
8 a

Plate 24

1. *Chlamys elegantissima* (Deshayes, 1863)
   17 mm; Japan, Okinawa, Seragaki
   (coll. HD) See p. 11

2. *Chlamys gilchristi* Sowerby, 1904
   15 mm; South Africa, Cape Agulhas
   (coll. HD) See p. 12

3. *Chlamys humilis* Sowerby, 1904
   15 mm; South Africa, Cape Agulhas
   (coll. HD) See p. 13

4. *Chlamys lowei* (Hertlein, 1935)
   13 mm; Galapagos Islands, Santa Cruz
   (coll. ZMA ex AR) See p. 15

4a. idem, 9 mm

5. *Chlamys multistriata* (Poli, 1795)
   19 mm; Italy, Sicily, Catania
   (coll. ZMA ex AR) See p. 16

5a. idem, 21 mm

6. *Chlamys princessae* Kuroda & Habe, 1971
   20 mm; Japan, Okinawa, Seragaki
   (coll. HD) See p. 17

7. *Chlamys quadrilirata* (Lischke, 1870)
   15 mm; Japan
   (coll. HD) See p. 17

8. *Chlamys weberi* Bavay, 1904
   16 mm; South Africa, Natal
   (coll. HD) See p. 21

9. *Chlamys wilhelminae* Bavay, 1904
   23 mm; Australia, QLD., Torres Strait
   (coll. ZMA ex AR) See p. 21

10. *Chlamys madreporarum* (Petit in Sowerby, 1842)
    18 mm; Australia, W.A., Pardoo Sands
    (coll. ZMA ex AR) See p. 23

1
2
3
4
4 a
5
5 a
6
7
8
9
10

Plate 25

1. *Chlamys inaequivalvis* (Sowerby, 1842)
   15 mm; New Guinea
   (coll. ZMA ex AR) See p. 43

2. *Chlamys camerella* (Berry, 1968)
   19 mm; USA, California, San Marcos Island
   (coll. HD) See p. 26

2a. idem, 20 mm

3. *Chlamys consociata* Smith, 1915
   18 mm; New Zealand
   (coll. HD) See p. 27

4. *Chlamys cookei* Dall, Bartsch & Rehder, 1938
   21 mm; USA, Hawaii, Oahu
   (coll. HD) See p. 32

4a. idem, 22 mm
   (coll. PP)

5. *Flexopecten flexuosus* (Poli, 1795)
   22 mm; Spain, Barcelona
   (coll. ZMA ex AR) See p. 40

6. *Flexopecten flexuosus duplicatus* (Locard, 1888)
   18 mm; Italy, Liguria
   (coll. ZMA ex AR) See p. 40

7. *Palliolum striatum* (Müller, 1776)
   16 mm; Norway, Bergen
   (coll. ZMA) See p. 48

7a. idem, 21 mm

1
2
2 a
3
4
4 a
5
6
7
7 a

Plate 26

1. *Lyropecten antillarum* (Recluz, 1853)
   24 mm; Netherlands Antilles, Curaçao
   (coll. PP) See p. 44

2. *Semipallium rapanense* (Bavay, 1905)
   24 mm; French Polynesia, Rapa Island
   (coll. HD) See p. 59

3. *Chlamys tincta* (Reeve, 1853)
   22 mm; South Africa, Komrie
   (coll. ZMA ex AR) See p. 20

3a. idem, 19 mm

4. *Chlamys sentis* (Reeve, 1853)
   30 mm; USA., Florida, Fort Lauderdale
   (coll. PP) See p. 30

4a. idem, 33 mm

4b. idem, 25 mm

5. *Semipallium amicum* (Smith, 1885)
   22 mm; New Caledonia, Noumea
   (coll. HD) See p. 58

6. *Juxtamusium coudeini* (Bavay, 1903)
   23 mm; New Caledonia
   (coll. HD) See p. 60

7. *Juxtamusium maldivense* (Smith, 1903)
   16 mm; Israël, Eilat
   (coll. HD) See p. 60

8. *Parvamussium alaskense* (Dall, 1871)
   24 mm; Japan, Hokkaido
   (coll. AH) See p. 67

1
2
3
3 a
4
4 a
4 b
5
6
7
8

Plate 27

1. *Propeamussium sibogai* (Dautzenberg & Bavay, 1904)
   42 mm; South Africa, Durban
   (coll. HD) See p. 66

1a. 28 mm; Japan, Wakayama
   (coll. AH)

2. *Parvamussium jeffreysii* (Smith, 1885)
   22 mm; Japan
   (coll. HD) See p. 68

3. *Cyclopecten vitreus* (Gmelin, 1791)
   17 mm; Norway
   (coll. HD) See p. 78

4. *Propeamussium dalli* (Smith, 1885)
   57 mm; USA Florida
   (coll. HD) See p. 64

5. *Propeamussium caducum* (Smith, 1885)
   23 mm (right valve, innerside); Japan
   (coll. HD) See p. 64

5a. 22 mm (innerside); Japan, Mikawa
   (coll. ZMA ex AR)

6. *Cyclopecten frigidus* (Jensen, 1903)
   20 mm; Arctica, Barents Sea
   (coll. HD) See p. 80

7. *Cyclopecten greenlandicus* (Sowerby, 1842)
   25 mm; Canada, Baffin Bay
   (coll. HD) See p. 77

8. *Cyclopecten vitreus macrocheiricola* (Habe, 1951)
   16 mm; Japan
   (coll. ZMA ex AR) See p. 79

1
1 a
2
3
8
4
5
5 a
6
7

Plate 28

1. *Bractechlamys evecta* Iredale, 1939
   35 mm; Australia, N-QLD., Great Barrier Reef
   (coll. AMS, paratype) See p. 8

2. *Chlamys perillustris* Iredale, 1925
   38 mm; Australia, Victoria, South of Gabo Isl.
   (coll. AMS) See p. 34

3. *Chlamys islandica* (Müller, 1776)
   49 mm; off Iceland
   (coll. ZMA) See p. 13

3a. idem, 80 mm; Norway, Lofoten
   (coll. JB)

4. *Chlamys tehuelcha* forma *madrynensis* (Lahille, 1906)
   40 mm; Argentina, off Puerto Madryn
   (coll. HD) See p. 19

5. *Mesopeplum triggi* (Cotton & Godfrey, 1938)
   41 mm; Australia, SA., Great Australian Bight, SW of Cape Adieu
   (coll. AMS) See p. 61

6. *Chlamys australis* (Sowerby, 1842)
   69 mm; Australia, WA. Shark Bay
   (coll. ZMA) See p. 27

6a. idem, 85 mm
   (coll. ZMA ex AR)

1
2
3
4
3 a
5
6
6 a

Plate 29

Nodipecten magnificus (Sowerby I, 1835)
60 mm; Galapagos, Fernandina Island
(coll. LACM) See p. 46

INDEX

The names of (sub)genera are printed in CAPITALS. Valid names of species and (sub)genera are in **bold.** Page numbers where genera and species are decribed are also printed in **bold.**